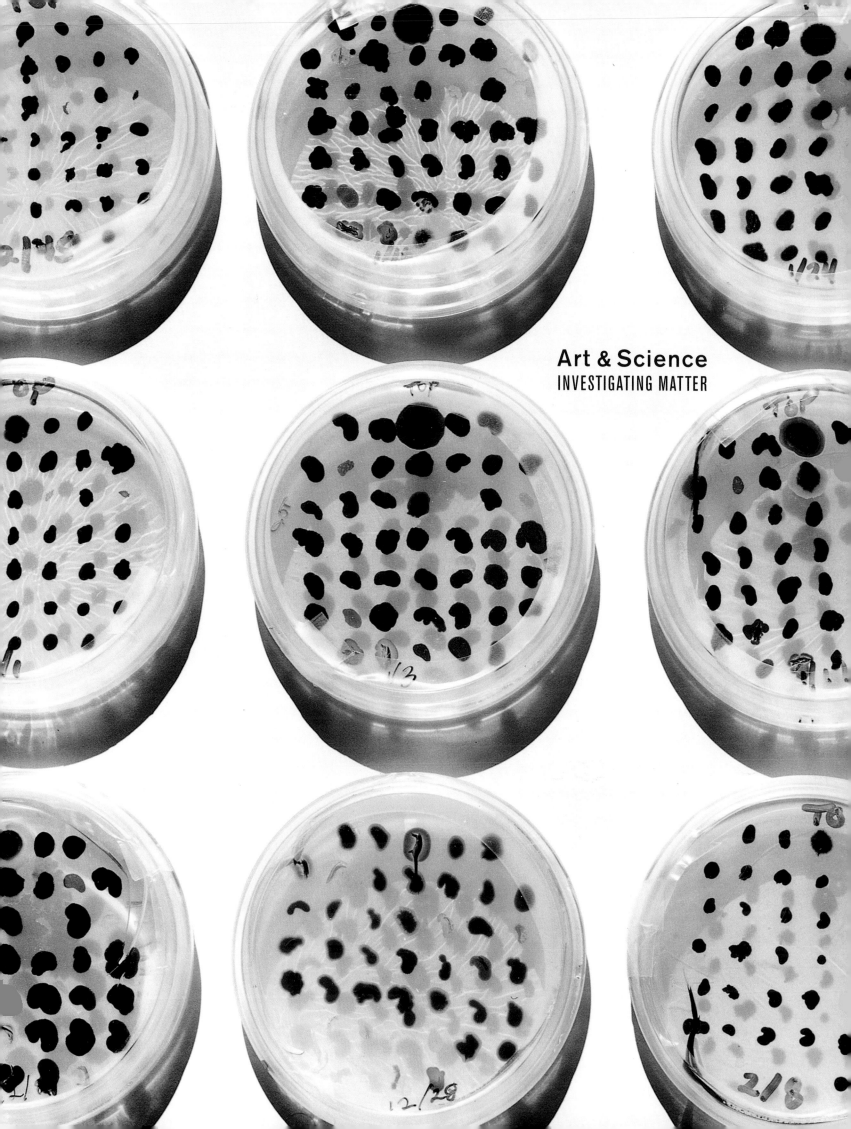

Art & Science
INVESTIGATING MATTER

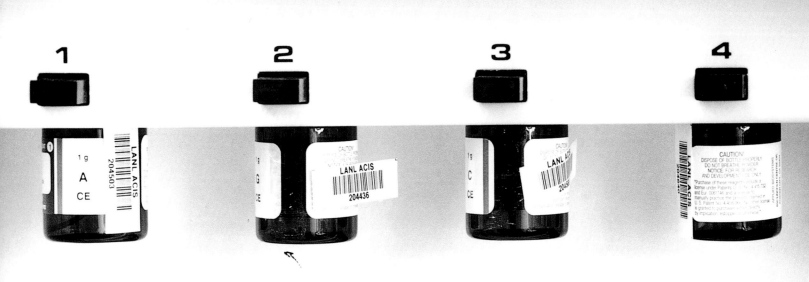

1

1 g
A
CE

LANL ACIS
204503

2

G
E

LANL ACIS
204436

3

C
CE

LANL ACIS
204

4

CAUTION!
DISPOSE OF BOTTLE PROPERLY
DO NOT BREATHE POWDER
NOTICE: FOR RESEARCH
AND DEVELOPMENT USE ONLY

9

LANL ACIS
196860

10

11

NTHESIS REAGENT **11**

180 mL

Synthesis Grade

tic Anhydride/
Lutidine/
rahydrofuran

400607

Applied
Biosystems

CATHERINE WAGNER
Art & Science
INVESTIGATING MATTER

Cornelia Homburg

with essays by

William H. Gass

Helen E. Longino

NAZRAELI PRESS

WASHINGTON UNIVERSITY GALLERY OF ART

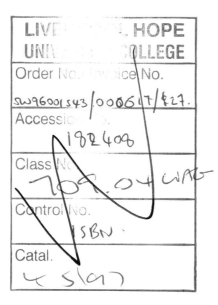
This catalog is published in conjunction with the exhibition *Art & Science: Investigating Matter*, organized by the Washington University Gallery of Art in St. Louis, Missouri.

Washington University Gallery of Art, St. Louis
September 6-November 3, 1996
International Center of Photography, New York
March 28-June 15, 1997

Support for the exhibition was provided by the Missouri Arts Council, the Regional Arts Commission, the St. Louis Printmarket and the Hortense Lewin Art Fund of Washington University.

Editor: Jane E. Neidhardt

Die Deutsche Bibliothek - CIP-Einheitsaufnahme
Catherine Wagner, art & science : investigating matter / Washington University Gallery of Art.
Cornelia Homburg. With essays by William H. Gass ; Helen E. Longino. - Munich : Nazraeli
Press, 1996
ISBN 3-923922-45-0
NE: Homburg, Cornelia; Gass, William H.; Longino, Helen E.; Wagner, Catherine [Ill.];
Washington University ‹Saint Louis, Mo.› / Gallery of Art; Art & science; art and science

Available in the United States through
Distributed Art Publishers
636 Broadway, New York, NY 10012

COVER:
Definitely Not Sterile, 1995
(cat. no. 8)

VERSO:
DNA/RNA Synthesizer, 1992
detail
(cat. no. 27)

BACK COVER:
Pipette Stand, 1995
(cat. no. 9)

Contents

Foreword

Catherine Wagner's latest body of work is an ambitious project that crosses the line traditionally separating the arts and the sciences. In the process of creating her art Catherine worked in a truly interdisciplinary manner with scientists, curators, educators, students and administrators. She entered laboratories across the country and constructed "still-lifes" from the subjects and objects of scientific investigation with a precision and clarity that brings the contents of the laboratory to light for the lay public. The scientists whose laboratories she visited welcomed her into their world and introduced her to their research, providing an intellectual exchange that generated new perspectives on the work of both. From this synergy Ms. Wagner created a body of photographs that not only exposes her questions regarding science, but simultaneously provides a means for scientists to step back and reconsider the context of their research.

The catalyst for Washington University's involvement with this project occurred in 1994 when Peter Marcus, Professor of Art, invited Ms. Wagner to create experimental lithographs in the School of Art's Collaborative Print Shop, which operates under the direction of Kevin Garber. While working on campus, she became aware of the extent of the University's research in biochemistry, molecular biology, physics, oncology, earth and planetary sciences and, perhaps at the leading edge of scientific research today, the human genome project. She then decided to pursue *Art and Science: Investigating Matter* here. Joe Deal, Dean of the Washington University School of Art, introduced her work to the Visual Arts Committee, under the chair of Associate Vice Chancellor Dr. Gerhild Williams, to endorse and support Catherine's proposal to spend a year at Washington University as an artist-in-residence. Under the sponsorship of Executive Vice Chancellor Dr. Edward Macias, the Louis D. Beaumont Fund of the School of Art and the Yalem Fund of the Gallery of Art, Catherine spent the 1994-95 year pursuing her *Art and Science*.

This publication and the accompanying exhibition present the results of Catherine Wagner's work to the public. As in the laboratory investigations, the catalog continues the collaborative, cross-disciplinary nature of the project. The authors for the catalog bring the perspective of their respective disciplines to this seminal topic. Dr. Cornelia Homburg, Curator at the Gallery, places Catherine's work in the context of the history of art and science and looks at how two such different fields have influenced each other over time. Dr. William Gass, David May Distinguished Professor of Humanities at Washington University and noted author, provides his literary reflections inspired by Catherine's photographs. And Dr. Helen Longino, Professor of the Philosophy of Science, University of Minnesota, reviews the economic, political and philosophical issues affecting science today and asks the difficult question, where will science lead us?

At the Gallery of Art, Connie Homburg, working closely with Catherine Wagner, was instrumental in the conception of the exhibition and catalog and dedicated the better part of two years to bringing this project to fruition. Jane Neidhardt, editor of the catalog, synthesized the three distinct essays and coordinated the production of the catalog. And the Gallery of Art staff, ever a dedicated and hardworking crew, made the exhibition possible. Finally, Chris Pichler of Nazraeli Press, Munich, demonstrated a devotion to the artistic concept of this publication that is rare among commercial publishers today. He and designer Robin Weiss are to be credited for the quality of the publication, while the collaborative efforts of John Randolph and Bruce Tomb of IOOA, San Francisco, led to the outstanding installation of the exhibition.

I must acknowledge Catherine Wagner for the creative vision to propose such a pioneering project that has brought together so many individuals and departments in the arts, humanities and sciences. Through her unique combination of artistic sensibility and intellectual rigor she has achieved what many have sought in our new age of cross-disciplinary studies. She has penetrated the inner sanctum of the scientist's laboratory and carefully staged each image to present for the public a visually stimulating view of its reclusive domain. Through her manipulation of scale, isolation of instruments, and assemblage of specimens—in combination with titles that reveal the stark reality of the subjects—her work offers us the opportunity to see into a hidden world and reflect on some fundamental issues of modern society.

JOSEPH D. KETNER
Director
Washington University Gallery of Art

Acknowledgements

My keen interest in photographing science experiments was born out of the work I completed for *American Classroom* in 1987. The children's classroom experiments I photographed seemed to act as a model for much larger life issues. *Seed Germination Experiment* (Ill. 2) depicts the life-death cycle while *Observing Skin's Protective Role* illustrates the delicate balance between preservation and decay. The last image in *American Classroom* is *Alfred University Science Classroom* (Ill. 1): dead frogs float in a container of clear liquid formaldehyde, waiting for the college biology class to begin the study of life. The work in *American Classroom* as well as my current work asks questions about the construction of contemporary culture. My long-standing interest in exploring cultural archetypes from the built environment (i.e., construction sites, classrooms and homes) has led me to the ultimate model of a culture under construction: science.

When I began reading about the human genome project, I was struck by the intent to determine a "genetic blueprint," first by mapping chromosomes and then by sequencing the entire gene structure of the human race. This project has gathered some of the most powerful minds in science to act as modern cartographers for our future. I am interested in what impact the changes that emerge from contemporary scientific research will have on our culture—socially, spiritually and physically. In my work I have tried to ask the kind of questions posed by philosophers, artists, ethicists, architects and social scientists. All of these questions revolve around one central idea: who are we, and who will we become?

My first photographs for this project were conceptual still lifes of evidence found in the laboratories at Los Alamos National Laboratory, Stanford Linear Accelerator, Lawrence Livermore and the University of California. I photographed the interiors of -86 degree freezers containing the archives of twelve areas of concern and crisis, such as alcoholism, Alzheimer's, bipolar disorder, breast cancer, DNA synthesis, HIV and research from the human genome project, to formulate the *-86 Degree Freezers* typology. In 1993 the Visual Arts Committee of Washington University in St. Louis accepted my proposal to continue my work using the University's cutting edge research laboratories. I gained access to the laboratories of molecular biology, conceptual physics, organic chemistry, psychology, biology, thermonuclear fission, physics, electrochemistry, human genome studies and earth and planetary science. In the earth and planetary science laboratory I was drawn to the fossils surveying life's history over millions of years. The resulting nine part typology is constructed of a compendium of life forms, including early sea life, plant life and mammals spanning the evolution of existence. While the fossil typology calls into question the sources of our past, the freezer typology beckons toward the new millennium.

1. *Alfred University Science Classroom*, in *American Classroom*, 1987

It forces us to ask, how, in the future, will we construct our individual and cultural identities?

The photographs in this book are meant to serve as a catalyst for thinking about the intersection of images and ideas. It has never been my intention to document science, but to stimulate discussion, ideas and questions about human existence. By posing questions that are inviting and inclusive of non-science audiences, I hope to provide a forum to reconsider science and its relation to us.

I am grateful to all of the scientists who invited me into their laboratories for my investigation. In addition, there are numerous people who facilitated my work in various ways. Mr. Joe Deal, Dean of the School of Art at Washington University, was instrumental in introducing my work to Washington University's Visual Arts Committee. Mr. Joseph Ketner, Director of the Washington University Gallery of Art, was a member of this committee and I am grateful for his commitment to the project. I am indebted to Mr. Deal, Mr. Ketner, Dr. Gerhild Williams and the other members of the committee who had the foresight to fund this project in its early stages. I would like to thank John P. Shaefer, President of Research Corporation, for supporting my work even though it is somewhat far afield from the main emphasis of his foundation. The Aaron Siskind Foundation, Mills College Faculty Research Committee and the Center for Photographic Art deserve thanks for awarding me fellowships to pursue this body of work. I am grateful to Howard Stein for his support. I would especially like to thank Dr. Helen Donis-Keller, one of the creative minds involved with the human genome project, for allowing me to photograph extensively in her laboratory and for her help in gaining access to other laboratories. I enjoyed our dialogue about art and science and feel that I have gained a friendship as well as an education. I am indebted to Dr. Steve Conradson, a physicist at the Los Alamos National Laboratory, who helped me gain access to the laboratories there when the project was in its infancy. Our extensive dialogue about the working methods of research scientists and the social ramifications of science stayed with me

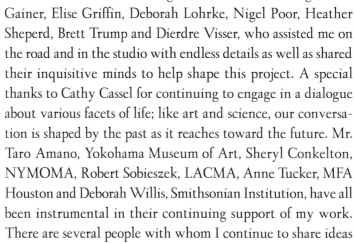

throughout the duration of my creative research. I am grateful to scientists Brad Jolliff and Dan Kremser for assisting me on the making of the *Moon Rock* typology with images made on the SEM. Scientist Laura Shawver of Sugen, Inc. offered enthusiasm and further laboratories to continue my work. A special thanks to Richard Messenger, who is both an artist and Vice Chancellor of Planning and Research at the University of California, Davis, for his assistance.

Further individuals I would like to acknowledge include Linda Armitage, Matt Gainer, Elise Griffin, Deborah Lohrke, Nigel Poor, Heather Sheperd, Brett Trump and Dierdre Visser, who assisted me on the road and in the studio with endless details as well as shared their inquisitive minds to help shape this project. A special thanks to Cathy Cassel for continuing to engage in a dialogue about various facets of life; like art and science, our conversation is shaped by the past as it reaches toward the future. Mr. Taro Amano, Yokohama Museum of Art, Sheryl Conkelton, NYMOMA, Robert Sobieszek, LACMA, Anne Tucker, MFA Houston and Deborah Willis, Smithsonian Institution, have all been instrumental in their continuing support of my work. There are several people with whom I continue to share ideas and to whom I continue to turn for various kinds of dialogues about art and life: Lewis Baltz, Frish Brandt, Anne Chamberlain, Amanda Doenitz, Danae Falliers, Jeffrey Fraenkel, Maya Ishiwata, Madeline Kahn, Hung Liu, Theresa Luisotti, Meridel Rubenstein and Leslie Smith. Simon Yuen deserves a gracious thanks for the many years we have worked together and collaborated on making exhibition prints. I feel that he has outdone himself in the magic he performs in the darkroom.

The exhibition and the catalog were made possible by the Washington University Gallery of Art. Douglas Burnham, John Randolph and Bruce Tomb at IOOA have shared in ongoing conversations about the conceptual interaction of architecture and art. I thank them for their involvement in the project since its beginning; we have become a collaborative team in designing the installation. Thanks to William Gass for his lyrical essay; it was generous of him to share his gift of language and seeing with my work. Respect and gratitude to Dr. Helen Longino for her philosophical essay that encompasses so many facets of understanding science and its social ramifications; I have long admired her ways of thinking. A sincere thanks to Dr. Cornelia Homburg, the curator of this exhibition. Her vision and commitment to this project were both inspirational and supportive. Her essay leads us through the fascinating history and marriage of art and science. I would like to thank Chris Pichler of Nazraeli Press, who undertook the publication of the catalog, for his commitment to produce a beautiful book. Jane Neidhardt contributed her excellent editorial skills to this publication. I am especially thankful to Robin Weiss, whose innovative design and thoughtfulness made the collaboration for the production of this book a joy.

And lastly I thank Loretta Gargan for her continued support. It has been more than a decade and we continue to share in rich conversations about the life cycle, and the questions about existence inherent in it.

CATHERINE WAGNER

ABOVE:
2. *Seed Germination Experiment*, in *American Classroom*, 1987

OPPOSITE:
Beating Heart—Heart Chamber, 1994
2 panel diptych
detail
(cat. no. 7)

FOLLOWING PAGES:
Fossils, 1995
9 panel typology
details
(cat. no. 3)

-86 Degree Freezers, 1995
12 panel typology
detail
(cat. no. 2)

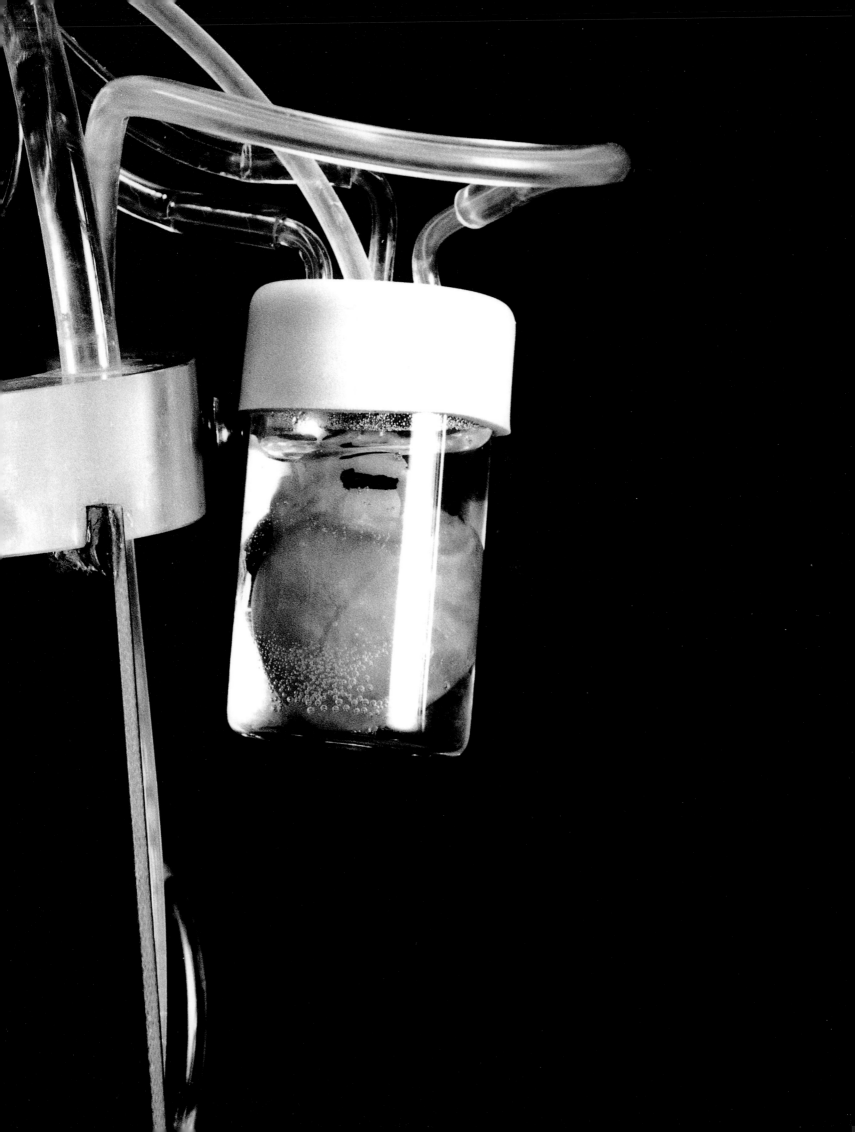

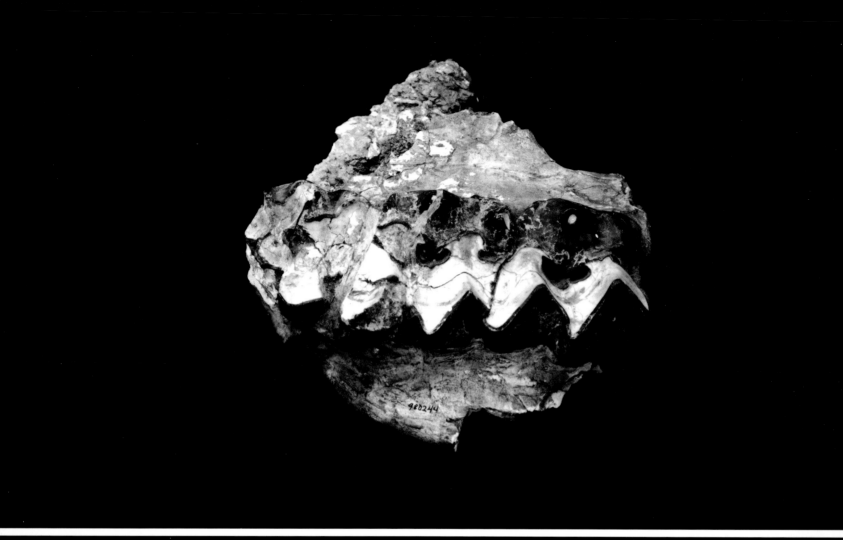

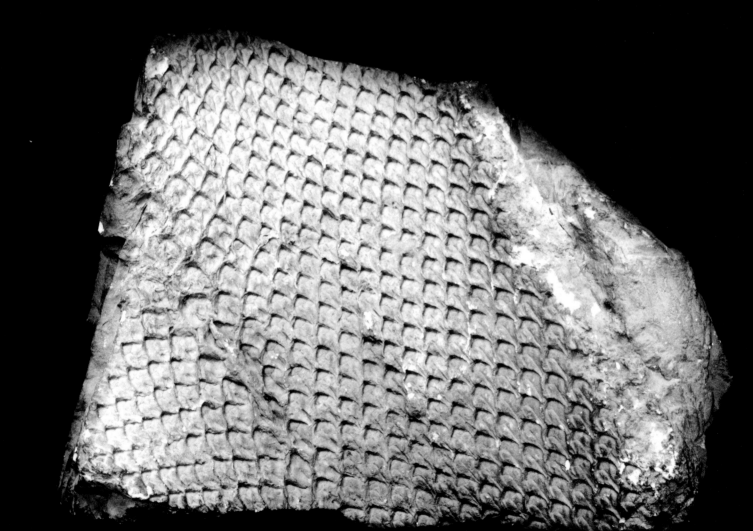

Art and Science

THEN AND NOW

Cornelia Homburg

Throughout history the results of scientific research have changed the human outlook on life and our understanding of nature. During the 20th century, however, the impact of science has achieved a prominence unknown before. Technical advances and scientific insights affect our awareness of the world more quickly and fundamentally than ever before, and, as our knowledge grows, so increases our ability to influence and manipulate nature.

An awareness of the decisive influence of science led photographer Catherine Wagner to undertake her visual investigation of this field. As an artist, not a scientist, she entered major research laboratories in the United States and constructed photographs of the world she found inside. She was drawn by a wide range of topics, from the analysis of moonrocks and fossils to medical research, in particular the human genome project with its incredible ramifications for humanity. Her resulting photographs offer the opportunity to encounter science in an innovative and unusual manner, as they not only cover various fields of research, such as biology, molecular biology, physics and earth and planetary science, but they also bridge the distance between art, science and everyday life. Wagner made her photographs as an inquiry into the overall concept of scientific research, at the same time creating images that live as works of art on their own.

While Wagner's approach to her project is a contemporary one that reflects the role science plays in our society today, it is part of a long tradition of artistic occupation with scientific research and its results. The relationship between art and science throughout history has been manifold, ranging from artists assuming the role of scientist, to the commission of portraits and works of art commemorating scientists and their achievements, to the use of science as a subject in art. Developments in the technical mastery of the arts have also been significantly influenced by corresponding developments in science.

In the modern world one of the most decisive impacts of science on art involved the mathematical calculations that led to the development of linear perspective. Invented in the early 15th century in Italy by Brunelleschi, an architect and painter, linear perspective rendered possible such a convincing representation of nature that soon artists all over Europe were eager to understand and utilize its principles.[1] The expressive possibilities of perspective have been explored throughout the centuries, with artists presenting carefully calculated schemata or developing elaborate illusionary spaces according to the preferences and ideas of different periods. Such expressions found their culmination in the Romantic movement, with its urge to convey the infinite space and grandeur of the universe. The British painter J.M.W. Turner, for instance, incorporated sweeping vistas into his composi-

Glove Box, 1993
(cat. no. 10)

tions, while his countryman, the painter and printmaker John Martin, was admired for daring perspectival constructions of extraordinary vastness.[2]

Just as the illusion of three-dimensional space significantly influenced the development of art since the Renaissance, a life-like depiction of the human figure became an essential part of artistic activity. Artists soon realized that knowledge about the internal structures and functions of the human body would help them achieve this goal and as a result were keenly interested in dissections of human bodies. While Renaissance artists had only occasionally the opportunity to watch or perform dissections, during the 17th and 18th centuries anatomical investigations became more accepted. Art academies provided opportunities for their students to attend dissections, and commonly used wax models of flayed bodies for instruction.

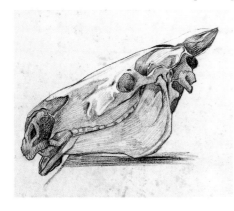

Artists interested in depicting animals also gained immense insight from and participated in such investigations. In France, for example, the Romantic painter Théodore Géricault made numerous anatomical drawings of horses for his equestrian compositions (Ill. 3), while the sculptor Antoine Louis Barye, who specialized in the representation of animals, and the painter Eugène Delacroix are known to have undertaken anatomical research themselves.[3]

If artists learned from science about the structure of a body, scientists in their turn benefitted from the skills and activities of artists. Frequently, scientific investigation raised public interest, and especially "anatomy lessons" received much attention. It was a special occasion when a famous scientist dissected a body in public, lecturing on the functions of various muscles, organs or parts of the brain. Artists were commissioned to depict such events in paintings or prints, among which Rembrandt's *Anatomy Lesson of Dr. Tulp* (1632) is one of the most famous examples (Ill. 4). While Rembrandt portrayed the highly respected surgeon conducting a dissection, the painting was not intended to represent a precise analysis of an anatomy lesson, but rather was commissioned as a celebratory portrait of the prominent man and a number of his colleagues.[4] Such visual commemorations of the achievements of eminent scientists were executed frequently, either by commission or by personal initiative, as in the case of Thomas Eakins' *Gross Clinic* of 1875. Other compositions depict the scientist surrounded by his work, in his study or in his laboratory, such as the portrait by Jacques La Joue of Abbé Nollet (1700-1770), who laid the foundation for our knowledge of osmosis and made important discoveries in the field of electricity, or of Louis Pasteur, who was represented in his laboratory by Albert Edelfelt in 1885 (Ill. 5).

In addition to commissions, artists also took it upon themselves to depict science and its tools in their work. Still lifes of medical equipment and instruments necessary for mathematical or astronomical calculations indicate the popularity of this topic. In many cases such compositions were also intended to invite reflection about the meaning of human existence. In a still life by P.G. van Roestraten (1630-1700), for example, a globe, a directional compass and two mathematical compasses are complemented by books and a skull, a combination that served to illustrate the erudition of the painter and at the same time places this painting in the tradition of the vanitas still life so popular at the time.

4. REMBRANDT VAN RIJN
*The Anatomy Lesson of Dr.
Nicolaas Tulp,* 1632

Artists also engaged in collaborations with scientists, who frequently employed painters or draughtsmen to illustrate their discoveries. A well-known textbook by Andreas Vesalius, *De Humani Corporis Fabrica* (1543), which for centuries served as a guideline for anatomical investigations, was illustrated by J.S. van Calcar, a pupil in the studio of Titian. Christopher Wren, who is best known for his architectural designs, made drawings of dissections conducted by the anatomist Thomas Willis and illustrated the latter's publication, *Cerebri Anatome* (1664). The numerous prints and drawings of plants, animals and parts of the human body that have survived over the centuries testify to the frequency of artists serving as scientific illustrators.[5]

New scientific data could also become an integral part of an artist's work, as in the late 19th century when Georges Seurat developed Neo-Impressionism in direct response to scientific discoveries about the properties of color. Seurat based many of his ideas on the knowledge he gained from such publications as *De la loi du contrast simultané des couleurs* by the French scientist Michel Chevreul and *Modern Chromatics* by the American physicist Ogden N. Rood. Convinced of the necessity and truth of his own conclusions, Seurat formulated his theory of painting and presented it on a grand scale to the public in 1886 by exhibiting his large composition *A Sunday Afternoon on the Island of La Grande Jatte* in Paris, where it received much attention and excited a following among avant-garde painters.[6]

The rise of photography in the 19th century resulted in new and unusual forms of interaction between artists and scientists. Developed through a scientific process, photography changed the visual representation of the world and profoundly influenced the making of art. It was soon employed by scientists and artists alike, who strove for both similar and, at the same time, very different goals.

In certain areas photography almost completely replaced artistic activity, as for example in the illustration of scientific publications. This function was largely substituted by the camera, which a scientist could generally operate himself. Photographs could be made relatively quickly, and they were highly favored because of the precision of the resulting image.

Both artists and scientists benefitted from the new possibilities of representation introduced through photography. Special techniques made it feasible to study natural phenomena that could not be detected with the naked eye. Eadweard Muybridge's groundbreaking studies documenting the different phases of motion are perhaps the most famous application of these advances. With the help of numerous cameras with specially designed shutters, he photographed animals and human figures moving in front of a gridlike backdrop (Ill. 6), and he was able to record phases of movement

which had been unknown before. These scientific insights attracted the interest of artists such as Thomas Eakins, a friend of Muybridge who made a number of motion photographs himself, and such painters as Edgar Degas, himself a photographer, and J.A. McNeill Whistler, who incorporated these new discoveries into their own work.[7]

The continuing rise of interest in science together with the industrial revolution propelled further achievements in scientific and technological areas, and these advances became apparent in the new urban landscape developing around the turn of the century. Numerous artists were fascinated by this evolution and chose it as a subject for their work. Charles Sheeler painted and photographed factories and other industrial structures in compositions that clearly indicate his interest in such technological achievements (Ill. 7). His contemporary Margaret Bourke-White shared his subject matter, but unlike Sheeler's her photographs tend to dramatize and glorify these accomplishments. Concurrent advances in the state of medical research also captured the attention of photographers. In 1926 the German Werner Mantz photographed an X-ray clinic, which at the time was still a novel sight, as Roentgen had discovered the technique only in 1895 (Ill. 8).

In the 1930s Berenice Abbott reacted to the increasing emphasis on science and technology and began to photograph scientific phenomena, a topic she would continue to work on for more than 20 years. She explained

5. ALBERT EDELFELT
Pasteur in his Laboratory, 1885

6. EADWEARD MUYBRIDGE
Animal Locomotion ("Daisy" Jumping a Hurdle, Saddled), 1885

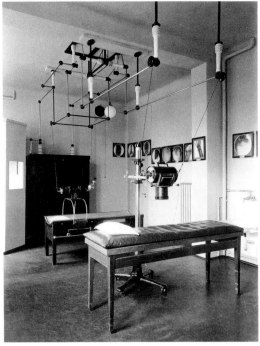

7. CHARLES SHEELER
Crosswalks, River Rouge Ford Plant,
1927

8. WERNER MANTZ
X-Ray Clinic, 1926

her motivation: "We live in a world made of science. But we—the millions of lay-men—do not understand or appreciate the knowledge which thus controls daily life."[8] Abbott was convinced that the precision of photography was particularly suited to visualize the immense influence of science, but she believed that the artist was better qualified than the scientist to convey this message to the public. She argued that the sensitivity of an artistic interpretation could function as a mediator between scientific fact and public understanding.

Abbott photographed scientific subjects initially on her own, and later in the context of an educational project. When she photographed a penicillin mold by enlarging it ten times, she captured its abstract beauty and dramatic force, but also emphasized its immense influence on human health (Ill. 9). As part of the Physics Science Study Committee of Educational Services, Inc., she produced photographs for high school textbooks explaining the basic laws of physics, such as the move-ment of waterwaves, the conservation of momentum or the motion of a pendulum. Frequently she developed new techniques to realize such images, which combine a didactic function with great aesthetic expressiveness.

Like Abbott, Catherine Wagner undertook her project of photographing science because she realized the fundamental impact of science on contemporary life. Wagner's work, however, does not attempt to document specific scientific experi-ments or explain physical laws; rather, it presents science more as a concept that reflects important goals and interests of our society. Wagner's photographs provide manifold views into the laboratory, focusing on its simple, auxiliary objects as well as its highly sophisticated instruments. Such detailed observations, however, are not used to construct a narrative, but to illuminate the essential characteristics of scien-tific work. As such they represent the intensity of research and its technical require-ments, but also the far-reaching and, for an outsider, awe-inspiring prospects of some experiments. Wagner invites her audience to appreciate the knowledge science generates about life, but also to realize the impact of this knowledge upon life.

Wagner's conceptual approach to her topic becomes apparent in the subject mat-ter she selected and avoided, the form in which she presented her imagery, and the level of involvement she allowed herself in arranging her objects. In order to avoid any interest in or identification with the individual, Wagner excluded people from her images and recorded only the results of human activity. She concentrated on the work of her fellow beings, but not their personalities or appearances. This principle is illustrated most forcefully in an image like the *Glove.Box* (cat. no. 10). Obviously, work has been done in the sterile interior that prevents contamination of our environment or protects a delicate experiment, but now the gloves that are thrust out at the beholder are empty. The somewhat startling frontality of the image emphasizes the absence of the person who has conducted research.

Wagner also chose to keep the locations of her photographs anonymous to avoid individualization. In this respect her work differs from other significant contemporary undertakings, such as the Siemens project or Lee Friedlander's photographic portrayal of the Cray computer company. Siemens, a large German firm that produces a variety of technical and electronic equipment, invited a number of photographers to interpret an aspect of the company, its activities and its people.[9] Friedlander was asked to document the working environment at Chippewa Falls, Minnesota, which is dominated by Cray.[10] In both instances the photographers created images that describe these particular environments, and only on a second level characterize the more general implications of living and working in and around such highly advanced technological and scientific sites. Catherine Wagner, on the other hand, was not interested in one specific place and, since she worked independently, had no restrictions to a particular site. As a result, her images do not identify any of the places where she photographed, but rather offer a more general representation of the essence of scientific research.

Wagner's black and white photographs stand out for their pristine quality and almost analytic clarity. The precision of her images is singularly suited to her topic as they seem to reflect some of the characteristics of scientific work: the analytic method necessary for research, the orderliness of the laboratory, the precise recording of experiments, the sterile, carefully controlled environment. These aspects of laboratory work are also echoed in Wagner's choice of images. The orderly rows of petri dishes with *Algae* (cat. no. 5) exemplify the repetitiveness of investigations that are executed over and over with ever so slight variations. Arbitrary piles of containers with fruit flies (cat. no. 19), one of the most important sources for genetic research, are kept to supply uncountable experiments. Equally, objects carrying a careful identification of their sterile or unsterilized state indicate a banal but vital part of laboratory work (cat. nos. 8, 21). On the other hand, the *Ultra High Vacuum Chamber* (cat. no. 11) has an almost futuristic appearance. Its gleaming robot-like

shape exemplifies the highly advanced equipment on which much of the success of scientific research depends.

This emphasis on the characteristics of laboratory work are reflected on another level by Wagner's compositional choices for her photographs. Just as scientific methods are categorical in nature, Wagner's images appear systematically organized in specific categories such as close-ups, still lifes and interiors. Close-ups focus on containers with *Intestinal Mount* (cat. no. 14) or on rows of *Bone Marrow Smears* (cat. no. 28). The latter are collected on individual glass slides with precise inscriptions of their origins. The uncountable samples indicate the vast number of experiments necessary to achieve results, but they also evoke a disturbing feeling that the individual (human?) source is of relatively little value or consequence. Other photographs are still lifes of instruments (cat. no. 9) or objects produced in the laboratory such as *Genetically Engineered Tomatoes* (cat. no. 17) fitted with barcodes that assure their reproducibility. Wagner also depicted interiors such as the *Glove Box* or, on a smaller scale, the inside of a sterilization oven in *Sterilized Pipettes* (cat. no. 24), in which the artist created an independent space by concentrating exclusively on the enclosed space of the machine. The metallic crispness of this photograph emphasizes the technical environment and underlines the necessary absence of human life.

It is important to remember that the clinical aspect of the laboratory—the anonymity of samples as well as the exclusion of the individual for the sake of general assessments and insights—is not a condition of fact, but is created and maintained by the individual scientist who undertakes the research. Similarly, the systematic categorization of Wagner's project reflects the organized nature of scientific research, and at the same time projects a sense of objectivity of the photographer towards her work. However, Wagner was deeply engaged in the creation of her images and the arrangement of her subject matter, just as the scientist is actively involved in what happens in the laboratory. In many cases Wagner constructed her compositions, or at least positioned her objects, according to the needs of her vision. When she photographed the petri dishes that held *Mating Reactions of Algae* (cat. no. 5), she placed them in strong sunlight so that their forms became double exposed, creating a visual effect that echoes the multiplying function of the experiment. Another time she became directly involved in the scientific process. For the typology *Moon Rock Found on Moon, Moon Rock Found on Earth* (cat. no. 4), the artist collaborated with scientists in order to produce her images. She inserted her film into their computer to photographically represent an x-ray, a topographical view, and a chemical analysis of moonrocks found in the human environment as well as in space. In this case the artist did not remain an observer, but took an active part in the scientific process. She allowed her work to blur the line between artistic contemplation and scientific function as the visually interesting images convey also precise information about the subject matter.

Despite her precise rendering of instruments and objects, Wagner is rarely concerned with a comprehensive presentation of the scientific material on which she focuses. She does not illustrate the activities of the laboratory and her photographs are always works of art instead of documentaries. Their aesthetic appeal is great, as they range from meticulously observed objects to almost abstract representations. At the same time the images evoke interest in the scientific research depicted—the precise usage of an instrument, or the nature of an experiment—and are balanced

Sterilized Pipettes, 1993 (cat. no. 24)

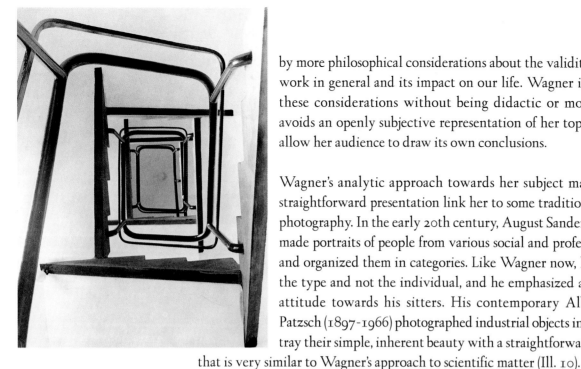

by more philosophical considerations about the validity of scientific work in general and its impact on our life. Wagner is able to raise these considerations without being didactic or moralizing. She avoids an openly subjective representation of her topic in order to allow her audience to draw its own conclusions.

Wagner's analytic approach towards her subject matter and her straightforward presentation link her to some traditions in German photography. In the early 20th century, August Sander (1876-1964) made portraits of people from various social and professional classes and organized them in categories. Like Wagner now, he focused on the type and not the individual, and he emphasized an impersonal attitude towards his sitters. His contemporary Albert Renger-Patzsch (1897-1966) photographed industrial objects in order to portray their simple, inherent beauty with a straightforward objectivity that is very similar to Wagner's approach to scientific matter (Ill. 10).

More contemporary parallels can be found in the work of the German photographers Hilla and Bernd Becher and some of their pupils. While a connection between their photographs and those by Wagner has been pointed out before,[11] only in this latest project has Wagner consciously established the link by assembling several of her images in larger groups and calling them *typologies* as the Bechers do. Similar to the Bechers' series of industrial structures, Wagner organized her laboratory photographs in regular rows and grids, evoking the idea of categories. The Bechers' images of water towers, blast-furnaces, gasworks and houses are characterized by a great uniformity of presentation which is achieved by excluding all references to weather and environment and their own involvement in taking the pho-

tographs (Ill. 11). While Wagner shares with the Bechers a conceptual approach to her visual material as well as the methodical arrangement of her photographs, her work differs in significant ways. To begin with, Wagner's topic is very different. She chose no industrial structures built by previous generations, but concentrated on an extremely advanced environment with high technology equipment. As such she does not assume the position of a chronicler, a role often found in the Bechers' work. Furthermore, while Wagner's typologies appear to deny a personal viewpoint, they nevertheless have a sensual quality which the Bechers carefully avoid.

This difference can be exemplified by Wagner's photos of the *Sequential Molecules* (cat. no. 1). The shimmering perfection of the flasks containing the molecules makes the viewer almost ignore their contents for the marvel of their beauty. This beauty is created by careful staging: the photographer used a smooth black backdrop and skillful lighting to

present each flask in its most alluring form. The bottles are arranged in profile or frontal position, recalling the tradition of portraiture in which the sitter is presented in his or her most advantageous position. The play of light gives the flasks a jewel-like quality. Wagner asked the scientists to inscribe their formulas on the bottles, thus describing their contents while also adorning them. The actual contents of the images are the molecules inside the flasks— fabricated by humans for use in experiments—a content that contrasts oddly with the beauty of their containers. It seems as if the molecules illustrate the control people have over nature, while their beautiful representation indicates the attraction that is connected to this manifestation of power.

This combination of methodical, categorizing observation with an evocative visualization characterizes a number of Wagner's photographs, and is used with great imagination to produce vastly different effects. The typologies especially are a strong case in point. If the flasks with molecules emanate shimmering beauty, the grid of twelve *Freezers* (cat. no. 2) produces chilling associations. The direct, frontal view does not allow for pleasurable contemplation; instead, the repetitiveness of the white, ice-crusted interiors is combined with the precise information of what they contain: materials related to research about Alzheimer's, HIV, cancer, human genome, etc. Visual presentation and content combine to confront the viewer with the implications of crucial medical research; projects that attempt to cure deadly diseases are closely linked to those that attempt to influence the genetic structure of the human body.

In direct contrast to the *Freezers*, a group of photographs depicting *Fossils* (cat. nos. 3, 6) creates an altogether different effect, recalling the long history of nature. While the freezer images are visually thrust upon the viewer, the fossils seem to float across their black background and therefore evoke a more mysterious, reflective mood. Rather than being reminiscent of past eras, the fossils provide a historical perspective for the contemporary human endeavors that are depicted in other images.

Wagner's use of a black background is not limited to the typologies, but appears also in a number of other photographs. In the *Beating Heart—Heart Chamber* (cat. no. 7), for example, bottles and instruments stand out strikingly against infinite blackness. Thin tubes are coiled inside of shimmering glass bottles on the right, while an apparatus on the left keeps the small heart artificially beating in its container. Wagner created a dramatic tension between the content, the rational scientific achievement, and its representation with the undefined background and brilliant, shimmering objects in the foreground.

The drama inherent in such images is balanced by those photographs that are more matter-of-fact. Such differences succeed due to the unifying objective of the project—a visualization of science as a reflection of contemporary interests and experiences. While specific photographs may highlight a particular aspect of scientific research, they are nevertheless part of a larger concept. For Wagner the importance of science today permeates all levels of society. Her images function as icons that testify to this overwhelming influence.

Wagner's sensitivity to the forces that shape modern life has been a thread throughout much of her previous work, but it reaches a new level of intensity here. Her involvement with her topic places this project in an historical context that links it to a long tradition. In this modern age of specialization, when it is hardly possible for an artist to assume the role of a scientist, Wagner has found a new way to extend the historical communication between the sciences and the arts, indicating with her photographs what can result from an interaction between the two. While her work reveals the world behind the laboratory doors, it is the objects, activities and results of these laboratories that inspired Catherine Wagner to create her fascinating vision.[12] ❖

OPPOSITE:
Mating Reactions of Algae, 1995
3 panel triptych
(cat. no. 5)

FOLLOWING PAGES:
Mating Reactions of Algae, 1995
3 panel triptych
detail
(cat. no. 5)

NOTES

1. For a detailed analysis of the development and use of perspective, see M. Kemp, *The Science of Art* (New Haven and London: Yale University Press, 1990).

2. See W. Feaver, *The Art of John Martin* (Oxford: Oxford University Press, 1975).

3. On the occasion of the death of a lion in the Jardin des Plantes in Paris, Delacroix and Barye met to dissect the lion and then made drawings of the animal after arranging it in different positions. See also *L'âme au corps: Arts et Sciences 1793-1993,* exh. cat. (Paris: Réunion des Musées Nationaux, Gallimard/Electa, 1993), p. 103.

4. For details on the tradition of commemorative painting and on this painting in particular, see W.S. Heckscher, *Rembrandt's Anatomy of Dr. Nicolaas Tulp* (New York: New York University Press, 1958).

5. While such illustrations are primarily considered as support for the scientific experiment or finding, they could also assist the researcher to clarify his own ideas about a structure or appearance.

6. For more about Seurat's contribution to art, see W.I. Homer, *Seurat and the Science of Painting* (Cambridge, Mass.: M.I.T. Press, 1964).

7. In the 20th century the study of motion was continued with the high-speed photography of Harold Edgerton. He developed the stroboscopic photograph which made it possible to visualize in detail any type of fast motion, ranging from a man hitting a golf ball to the flight of a bullet. The technical advances in photography have, of course, continued throughout the 20th century; scientists are now able to use extremely advanced computer equipment to record any type of subject matter. See H.E. Edgerton and J.R. Killian, Jr., *Flash! Seeing the Unseen by Ultra High-Speed Photography* (Boston: Hale, Cushman and Flint, 1939), and H. Robin, *The Scientific Image* (New York: Harry N. Abrams, Inc., 1992).

8. Berenice Abbott, from an unpublished statement in 1939, reproduced in *Berenice Abbott, Photographer: A Modern Vision,* exh. cat. (New York: The New York Public Library, 1989), p. 58.

9. See Thomas Weski, ed., *Siemens Photographic Project, 1987-1992* (Berlin: Ernst und Sohn Verlag, 1993).

10. See Lee Friedlander, *Cray at Chippewa Falls* (Minneapolis: Cray Research, Inc., 1987).

11. See Sheryl Conkelton, *Home and Other Stories: Photographs by Catherine Wagner* (Los Angeles and Albuquerque: Los Angeles County Museum and University of New Mexico Press, 1993), pp. 5ff.

12. The contemporary significance of Wagner's project is also reflected in recent moves by other artists to focus on advanced scientific achievements in their art. In England, for example, Peter Fraser has undertaken a series of photographs that depict instruments of the laboratory. He is also one of a group of artists invited to create artworks that reflect their ideas about the leading laboratory of genome research in Great Britain.

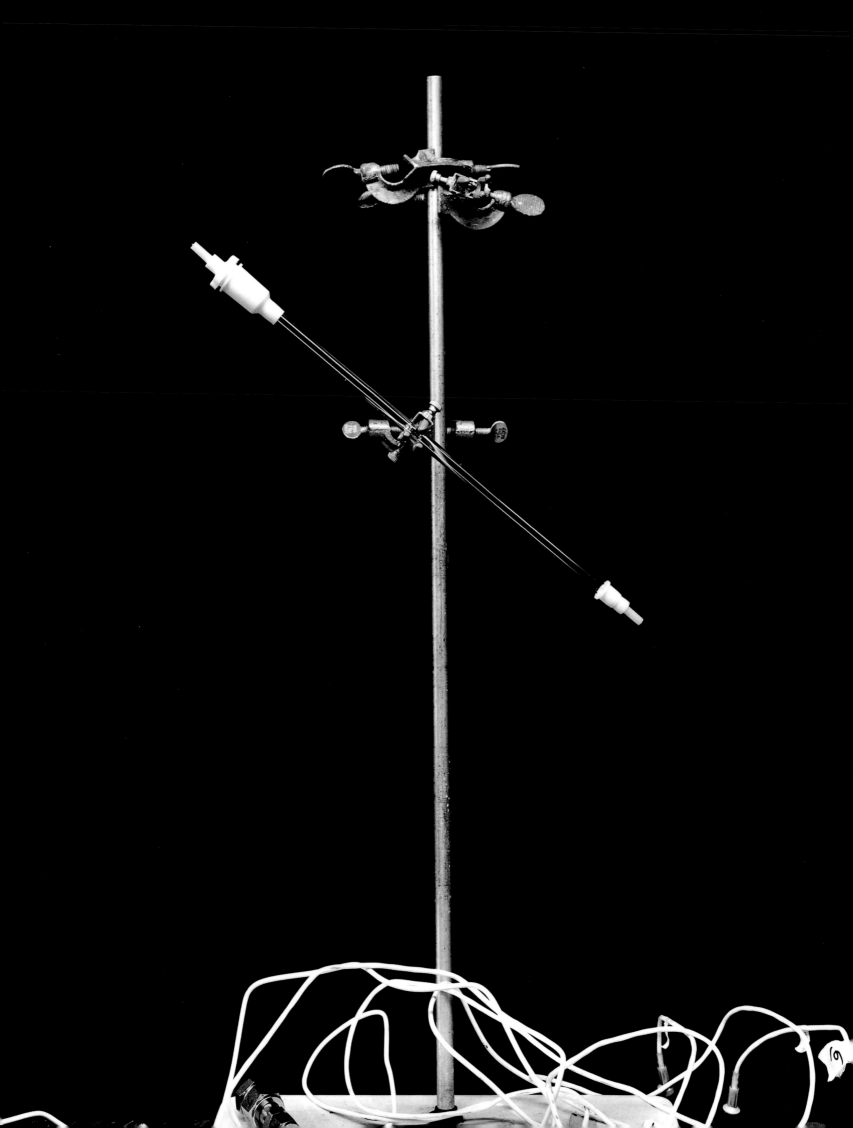

Catherine Wagner's

LABORATORIES

William H. Gass

Seeing is so important in science that the task is seldom left to the unaided eye, and never to an untrained one. Scales weigh and then say what they've measured. What is said is then seen and recorded, or recorded automatically and later registered in a brain. The scientist may be looking at something because society has a problem with it: perhaps it is seeking a cure for AIDS or polio or cancer or heart disease, and has urged with its interest, and encouraged with its funds, the mind-eye instrument, the hand-tool skill of the neurologist, the chemist, the modeler, the surgeon. However, the laboratory technician is looking, when she looks, along a long line of causes, reading surface for the sake of depth. She is seeing neither the petri dish nor its unappetizing jellies, but growth and disturbance, cause and effect, behavior and meaning. Traditionally, the scientist is indifferent to matters of public importance, because social interests are not automatically a form of knowledge, although the consequences of what is learned may be enormous: decoding, classifying, splicing, contriving DNA, for instance. Thus the democratic indifference of the instrument is to be treasured; the beaker's callous sides, the tube's willingness to bend, the microscope's enlargements are simply there, without prejudice or pride, to perform and serve.

Labs are not dogs, then, whose backs must be patted and rumps rubbed. But they are engines of education, and must be kept as much in trim as any rocket whose failure may injure many and cost millions. So the scientist will detect in her surroundings any inadequacies which could interfere with her project, her search. She guards against equipment failure, worries that an experiment may be compromised by slovenliness or impurity; and eventually, if all goes well, she discovers at the end of an eyepiece, perhaps, one of the secrets of nature, as those who hang around outside science's circle of competence are pleased to describe it.

The purity and crispness of Catherine Wagner's photographs are therefore quite appropriate to their present subject. They are objective in every sense which science admires, wholly given over to scrutiny, and to the things her lens loves. They see without the clouding of the eye. They do not personify. They do not pretend to be peeking at spaces which are fast asleep or at tools which lead another life during the night. The syringe does not dance in the dark, the freezer simply is, colder inside than any climate, vials vessel even when unobserved, a hose is never a snake, to hiss where it's hollow. Such is the immensity of her respect.

Pipette Stand, 1995
(cat. no. 9)

Centrifuge, beaker, test tube, vacuum chamber, scalpel, rubber apron, gloves, a freezer chest, a sink: servants without souls, without families to feed, roofs to uphold, fashioned for a function like a lock or key (though not the fruit fly, presumably); here a table, tray, or there a chair, a set of steel shelves, cords and cables, lamps like pitiless interrogators, dials designed down to the last dot to do one thing—point out. These and all the other instruments which staff a lab are altered altogether when they become ends in themselves, just the way wastebaskets, blackboards, books, the teacher's desk top, rows of chairs, trays of chalk did, waiting in the places Catherine Wagner photographed them in her equally powerful previous collection, *American Classroom*. The population of these spaces are now things with their own contours, textures, claims. They, not their employers, are front and center for a change. They are not simply empty vials, dead bugs awaiting disposal, or geneticized tomatoes, tagged like newborns, but comprise a glove box with its own armless thrust, glass with its glimmer, bottles of blooddark radioactive life connected by white veins to an unseen ceiling, a funnel like a wide Y, frost as granular as sugar and as protective as skin. The drosophila morgue, for instance, stands on a white table backed by wire mesh. The jar might contain thistle seeds. Inside the glass are shadows, dark inverted hills above a thousand bodies, and the shimmer of the mesh, curving too, is alive with light that's fallen through the funnel. Innumerable, the dark marks of the mesh, the seedlike black bodies of the fly; dazzling, the funnel's white void, and the table's glistening top.

The clone library lives in the deep freeze, a chest which holds a simple stack of labeled trays and boxes. Nothing else to see but the frost-knobbed doors, plain yet obscure words and numbers: "pooling source," "YH," or "10/14/92," and then, looking closely, the paw print of an animal caught in the snow or a patch of some previous thaw in the shape now of submerged tears. Well, these are not the spores of a polar bear or arctic fox, certainly, but indents just the same, and significant, since even a tire tread can catch a hit and run. They say: look again.

So sober a cave to contain such surprises, packaged like reams of paper, DNA by the dozen, frozen brood, I think, standing in front of the photograph, feeling the cool air escaping toward my chest. Open yet another ice box: O P Q R. . . faintly, S T U V. . . faintly, W X Y Z: labels in the letters of the alphabet for breast cancer research. V for Vera, I wonder, O for Olive, U for. . . for Undine. Below I read "Jackie L—cell." Jackie? Rows of small tubes, small jars lying on their cold sides, small boxes, antarctic remains, a camp abandoned during a dash to the Pole: here it is minus 71°.

Slide after slide of bone marrow smears, looking like scales taken from a huge fish, and cryptically called "27536 BM 4M 8/10/93 PAD 03" (not an E-mail address), lean on discreet rails in another image; though these are not images exactly, because they are simultaneously too much the pictured things themselves, and too much the picture. Through the glass the rails slide like shadows, another kind of sample taken from another world. The numbers are suddenly somewhat menacing.

The mind makes a memory theater of the eye, even when it stares emptily into a sterilizer where rocket resemblances arise quite unbidden by the rack in which the pipettes rest, so if I turn my head to right the ruled lines, and shake off the simile, I see a building made of modern towers. In the lattice lab where particles having less than any size are somehow confined I observe cotton candy rising like white smoke from crimped crusts which cover pie pans made of football leather. Unbidden, they must be banished, these fantasies, for they interfere with the immense presences even clear plastic packages contain: coils of dispensing tubes and triangles of aluminum foil, marked like wines with their years—"94," "93." No.

Unlike wines with their years. Patches of light reflected from the plastic, caught in mid-flare, float firmly above the foil, above the mountainous terrains of Utah or Wyoming, empty of life, seen from a plane. No. Unlike the moon's land, I remind myself, rather just like rumpled foil, riven by shadow, the way the storage sack is ignited by light.

Most of us will not recognize these objects or their intended actions, and only vaguely comprehend the spaces where they comport themselves. The foamy cotton batting, the foils, the dials, the dark liquids, the ruled tubes, those dark medicinal-looking bottles are largely strangers to us, and while our ignorance may allow us more innocently to see, it also encourages our thoughts to seek comforting comparisons—comforting only because they bring these scenes and their players into contact with concepts and experiences we understand.

Examining the photographs collected in *American Classroom*, it is not difficult to see how Catherine Wagner arrived at her present interest, because quite a few of the classrooms she visited before were laboratories or like them: a table covered with plastic cups in an elementary school which was the site of a seed germination experiment, or a collection of apples left to wither (Ill. 12) so that sixth graders could begin to understand the skin's protective function (a photograph which closely resembles the tagged genetic tomatoes of the present exhibit), or a stunning tableau made of frogs awaiting dissection at amateur hands, or molecular models in a science lab, even a milking machine, a loom in an institute of textile technology, as well as pressure gauges at MIT or pieces of a transmission entabled at a trade school. One set of tools is used to teach what's known, the other set to discover what isn't. The classroom has a hierarchy the laboratory hasn't. In a lab vials are corked, tubes sealed, ice chests closed, bottles are swaddled, whereas schoolbooks are open, specimens are on exhibit, maps drawn down, blackboards scrawled. The camera turns warm classroom woods into somber grays, but black and white labs remain black and white and are glad of it.

Order in the classroom is social and instantly visible; the laboratory order is hidden within purposes the observer does not normally know—the neat, the tidy, the sterile replace ranked rows of chairs, the teacher's desk, the lectern. Minds in the classroom are exposed, but the laboratory is a world of cordoned off countries, of artificial atmospheres, of extremes of heat and cold, of small and smaller and smallest. In the classroom no one should fear or come to harm, but labs are perilous places; rats, mice, frogs, flies die; poisons percolate through tubing, acids gnaw helplessly at the glass of their jars, gloves are worn, rubber aprons, masks; there are many soaps and sinks. Open me, the classroom cries out to Alice. Keep me closed, the laboratory door says firmly. No unauthorized admittance.

There are no people in these pictures. The spaces they depict are empty of such distractions. People are as unkempt as grass and solicit our sympathy instead of

12. *Observing Skin's Protective Role*, in *American Classroom*, 1987

OPPOSITE:
Solvent Stills, 1994
(cat. no. 15)

revealing their beauty. And we would see the children not the chairs, the technician's hands not the curvaceous vessels. The tools would become merely tools, losing much of their Being by being only what is done with them. The blackboard whose explanatory notes and diagrams now speak to no one, words which await erasure, lessons presumably learned, they are now like the poems no one reads; they address the void. "Caution!" a bottle warns an empty room. Yet we read "Caution!" and cannot reach the bottle whose label thus addresses us. "Caution!" does not say "Caution!" in our case; it simply is the word.

Immanuel Kant argued that the esthetic experience was (a) not mediated by concepts, (b) marked by a condition of disinterested interest, and that (c) its object exhibited purposiveness without purpose. Of course I may ask "what is that?" and lose my esthetic distance when told it is a jar of dead fruit flies; I may have had my hand many times inside the glove box, and felt its protective cover on my skin; but I must not miss what Catherine Wagner's photographs show us: the purposive machines (freed of function but not of form), the row of button-like rivets which fasten the glass and seal the opening (but frame the scene), the range of sterile whites and grays, the distant glisten of containers, syringes white as doll house columns, the black knuckles of the gloves, wall bolts, pipes, a reflection hiding in the window like a ghost mouse in a painted hole.

It is allowable to look at the stacked stock of drosophila and wonder at life itself, see mankind's time as short as a gnat's when set against the sun, even to imagine a minuscule consciousness within each dot, and write, as a poet might, unread words, though eloquent, upon the subject; yet it would be a shame if misguided pity led the eye inward at the wrong time, so that the position of the eight capped cups was unenjoyed, or the disposition of the specks was unappreciated. Alas, art is a lot like science, and its passions are dispassionate, its concerns elemental, it observations detached as the button that was lost last week from the sleeve of your winter coat.

A writer without rhythm is surely a wretched writer, for woeful is he who cannot give some music to his meanings. In all the arts, balances must be followed by graceful falls, and then by restorations. Life is similarly repetitive, routine, performed in habit's habit, but then varied always, sometimes insidiously, often as obviously as a sneeze in church, a yawn in front of company; and the camera catches the pattern, it allows its objects, however fortuitously they may have come together, to feel they are in a community of Friends, silent and peaceful, dressed in consensus. It unites steady circumstance with chance, discovers a fresh whole, and lifts, like a smear upon a slide, that prized bit of the visual, and slips it inside a frame the way those two sails we saw were placed under water-looking plastic where the rolled tubing outlandishly forms a wheel, alive with light as if the wheels were burners. . . . There I go—up like smoke into similitude. No. Tubes are tubes, coiled perhaps, but still unsterilized because the caption says so, and the sails are simply sheets of foil wrapped around tube ends, and each dunked in a baggie almost like the kind you buy to store food. We have stood before this photo before, but that is what happens when, instead of passing through the picture to what's pictured by it, and then, through that, into the world the image is an image of, we remain within the photograph like one of those jars in another photograph, arranged in royal row, dark as bodies against the white wall of the DNA synthesizer, eleven labeled eleven— just so—in order that we may realize, experience, thrill to the power of these depicted relations.

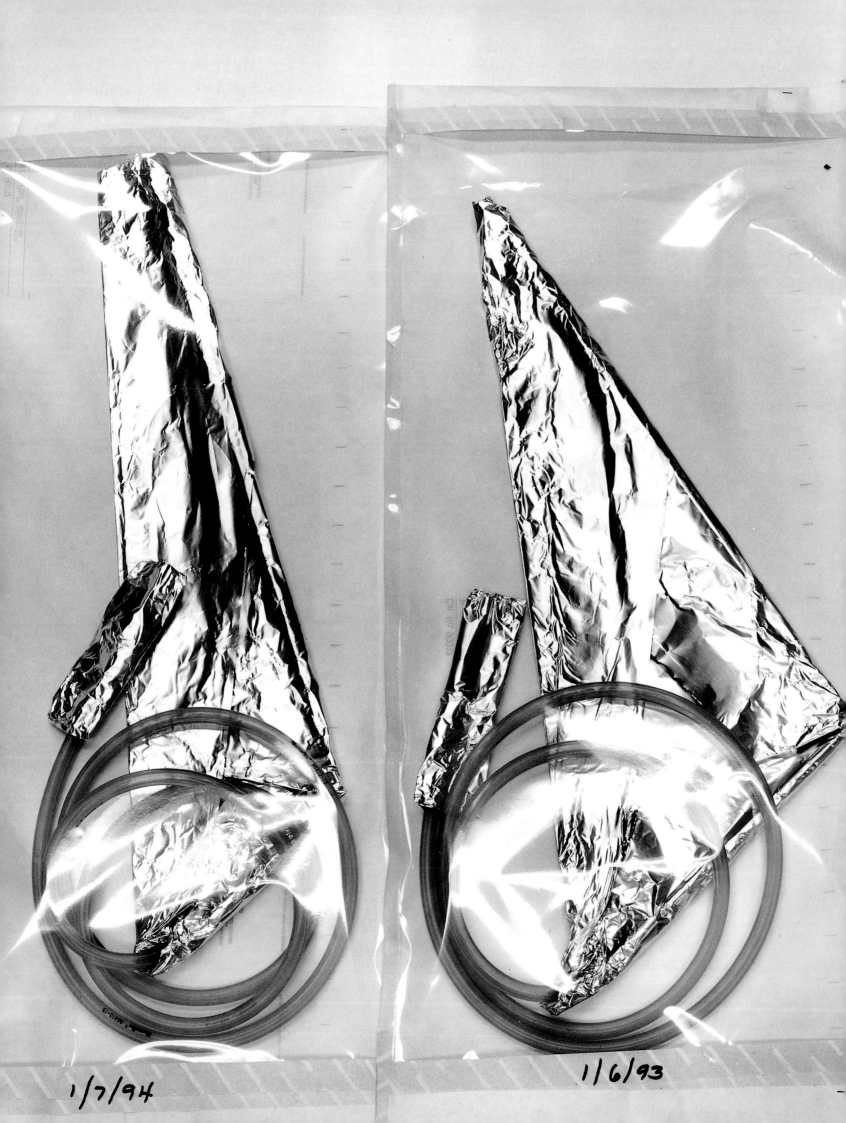

1/7/94

1/6/93

Catherine Wagner's art (as in all art, to my mind) seizes upon the connections it sees, freezes them for an indeterminate future, exhibits the beauty that inhabits schoolroom, lab, or nature, the splendid company which things in themselves have, all in line like those jars from one to eight, the dark buttons above, the dark dial to the right, the dark tops to the flasks at the upper left, and then the procession of large dark bottles suspended from their necks like executed enemies—not standing, hanging—the black numerals naming them all the way to fifteen, and finally the dark end of the marking pen at a cunning corner of the composition. Rows and ranks everywhere, similar shapes, identical entities, a keyboard, rectangularities of every kind, registering, by being white, the rolling and the round of the bottles and the jars, perversely lit from within by reflection in spite of being blank without their tags.

Unsterilized Dispensing Tubes, 1994 (cat. no. 21)

In the early days of photography, you sat solemnly for your photograph in carefully controlled surroundings. There was perhaps a chair, a drape, a pasteboard tree, a painted sky. You wore the dress or suit you'd be buried in, and, motionless for many moments, stared into the camera as though mesmerized. The photographer, his head hidden in a black sack, held a pan of flash powder aloft, and there was something near for you to look at—called the birdie—but the camera's lens was pointed at you like the barrel of a pistol. A puff of smoke, perhaps (has the gun gone off?), a brief brightness gives your cheeks a pat, and then your image slides magically down a tube to rest inside a plate of glass where your form and face will be preserved for your family or the files of the town's principal paper to remind the world who you were. Showing no more animation than the living were allowed, the dead were similarly caught, dressed for a party even in their coffins.

Catherine Wagner has arranged for nine flasks of molecules to pose for her, each beaker on its own cylinder of cork, but draped in darkness as if to reverse old-fashioned procedures, and so cunningly lit light seems nowhere else than in the molecules themselves. They glow like unsocketed light bulbs would if the bulbs could hoard a little of their ordinary burning to read books by when switched off. Beaked like birds or stoppered like decanters, and glowing like jewels on a velvet covered counter, there they are: *dasein* could have no better referent, and every *likeness* does them harm. Yet one is snow-filled, another has cloud.

We know molecules are movement, so the gleam which comes from their containers seems appropriate. How differently posed the fossils are, made of molecules, too, but petrified and at a standstill here in a different darkness—the darkness of the past from which they seem to have emerged, teeth caught in a jaw, toes or talons like roots but lacking either bird or tree, fragments, as every photo is, of creatures in a world more vast than wide-angles can attend to. The paleontologist will relate these shards to the pottery God blew his breath in. He can connect the skeleton together in his mind. The plated skin, scales pretending to be tire treads, countless creatures dead together in a perfectly incised design: for him any one may be a puzzle's lost piece, perhaps, though no doubt these particular fossils are exhibits only, standing for a kind—the sort of thing, in ice or cave slime, in slate or limestone rock, we find.

Fossils, already as firmly defined and settled in themselves as photos, find their forms fixed once again on film. It is ironic that these fractions should be given an attention appropriate to a healthy whole, for they *are* wholes now, whole bits, and have the bitter Being of the total fragment to hold on to, even when shown to students as representatives of a type, a clan, a class. It would not do for them to hide behind a shower of glints; they must show their edges and their mars and wrinkles; fine details are all they are; their history lies in line and lump and ridge. In the environment which Catherine Wagner has devised for them, these remnants are as particular as they once were when fully-jawed and chewing.

The scale of the school must reflect the size of the student, the child rather than the teacher, although whatever is generally human may serve as the *modulor*. In the observatory or the lab, however, there are many suits and sizes: XL often, for hot and cold, M for the ceiling, walls and floor, XS for molecules, for genes and XXXS in the nano realms. Telescopes bring near the unimaginably Vast but from such incomprehensible distances the Vast remains the Large seen small, nebula like a leak of cream in coffee, while microscopes inflate microbes to a mouse's measure and draw up other things to view that are so tiny they are no longer things, not even elements, not even motes in a needle's eye, so small they've had to be invented.

The path of an electron can be caught on film, and depths can be plumbed which reach so far beneath whatever is beneath us that the treasure found must seem like a buried star; but Catherine Wagner's camera, as supervisory as it is, does not peer through any other instrument. It is satisfied to stay in *modulor* range of its objects. However, that great wrapped Polyphemus, the vacuum chamber, is downsized to fit in her 4 x 5, then enlarged for our looking (all these monkeyshines with measurement don't seem to bother most people, who find nothing odd in seeing a great sun like a speck in space, collecting it on a minuscule stretch of film, only to make the filmed dot greater than a dinner plate; but I worry a lot, and I wonder what I'm looking at, at any time, and how hard on Alice's neck it was to pole up that sooty Wonderland chimney). (If I were eight feet tall, and wore a top hat, would my thoughts stay small? would I be 'I' at that height? and if I were a giant how much would you pay to see me in the sideshow? dime for every inch? too high a price for such a teenyweeny tall guy? and if I were a giant would I step on your head with my 30 size shoe?)

"Of course," the answer comes, "if *you* were eight feet tall *you* would be no longer who you are, but if your image is enlarged, *you* are wholly unaffected." Images can be piled on top of one another like toy blocks, images can be shrunk into the dainty oval of a brooch, or blown up like balloons to be bigger than a billboard. I can carry a picture of you in my wallet; I can prop a picture of you and the kids on my office desk; but I can't change the scale of a Rembrandt without esthetic alteration, and a photograph can't be resized without producing an importantly different effect. There is probably, for every photograph, the right dimension. Life size may be the most disconcerting. Images, like signs, are most often made to replace their referents with more easily managed replicas. Ten thousand slides can fit a city into a single cabinet. *Pictures* belong in books, in trays, on gray screens, in billfolds, on bedside tables, above sofas. *Photographs* are their own boss, and achieve, as these do, an independent being. A tower built of freezers is not a freezer, nor is their pictorial amassment merely a multiple of one. Both are fresh things, imposing, possibly imperial.

How can a photograph tell us something about its subject matter, then, and stay a photograph? Because, I would reply, every photograph is a picture first. Examine one. The freezer's door stands open to disclose a frost covered window—so it seems. Beneath the door is a dial and next to that dial, just beyond a bunch of switches and buttons, is the decal of an American flag. Is it possible for me to read that emblem with a wide smile, to note the contents of the freezer and turn them into a visual simile, a winter window, while, at the same time, in the same look, with a single, and the same, pair of eyes, to appreciate pure layout—freezer form and moisture texture? Not only do we do it—we all do it easily—but we enjoy the tension as well as the harmony between these values: the real freezer is a fragment, here, but a pictorial whole; the flag is a small wry joke, but an important part of a subordinate composition; the freezer's contents are as frozen as fossils, but the window pretends that it is cold outdoors.

And how can a photograph do justice to all its elements without singling out an object, surrounding it with the void, and taking its picture on a dais like a divinity, or at least the speaker for the evening? The instruments of the laboratory are subordinate to their investigative ends, the flag in the corner is a soft note in a large composition. Art also hails hierarchy like a cab. But in the lab the fruit flies die, the vials die, the freezers finally fail to maintain their hibernating cold and are drawn away in the back of a truck to be junked, tubes need to be replaced, beakers break, little plastic boxes crack, even the technicians come and go, grants are not awarded, research takes a different turn, entire experiments go as sour as limes. In the photograph, however. . . is realized—to return to Kant—a true community of ends; for the ideal of art, I think, is to give each dimension of the medium its due, its full and fair share: matter, mind, imagination—trope, thought, thing— function, form, feeling—use, design, dream—percept, concept, precept—theory, fact, and fiction; and to allow the various elements of the composition, whether emblem, bolt, or bottle, an uncoerced allegiance to the whole.

In the laboratory, as well as in its *picture*, corners come together on account of the cabinet, bolts put bars in place, *this* phenomenon requires *that* treatment, mechanical laws can brutally compel obedience; but in the *photograph* alignments made necessary by nature are read as free, because things are where they *ought* to be. When the Demiurge, in Plato's *Timaeus*, creates the world of appearance, he can make some of what he does come about purely in accordance with reason, other things remain recalcitrant (the qualitative character of qualities, for instance), while in still other cases, reason is able to *persuade* necessity: that is, to make the blindly necessary something one would have rationally chosen. Much of the art of the photograph depends on this ability, for the camera has, and is, "a finder." It turns accidents into ends; it pulls orderly groups out of chaotic heaps; it rerelates things in "picture space"; it forms by finding an angle on an object, by raising a simple little shadow to royalty, transfiguring things by highlighting their details, through scale, framing, paper, focus, putting all things in positions which respect their integrity and treat them as ends in themselves, but in stations of this new society whose duties can be freely taken up, because, in the Kantian kingdom of ends, each agent rules only her own will—by willing what would be appropriate, enhancing and liberating for all.

In Catherine Wagner's photographs, the freezer doors assent to the swing of their hinges, the frost lies down on the cabinet's metal like wild grass its meadow, but only in the poem she's made of this appliance. Nor does the art of her eye, as calm yet active as the rooms it reveals, allow us to forget the wonders the case encloses, or the encoded life the cold holds. ❖

FOLLOWING PAGES:
Sequential Molecules, 1995
9 panel typology
details
(cat. no. 1)

Science in Context

PHILOSOPHICAL REFLECTIONS ON LABORATORY SCIENCE

Helen E. Longino

The scientific laboratory resonates with the hopes and anxieties of modern industrial societies—hopes and anxieties about time, space and ourselves. We seek from fragments of the past clues to what we've been and what we are now, we seek from fragments of the present clues to what we will become; we seek from images accessible to ordinary vision insight into the vanishingly small and the vanishingly large—from subatomic particles that are the ultimate constituents of the material world to galaxies and gasses at the edge of the cosmos. In the laboratory fragments of the world are dissected, isolated, recombined, weighed and watched in order to create accounts of their operation.

Not all scientific investigation takes place in the laboratory. Theoretical work involves articulating general or encompassing relationships among phenomena: finding mathematical structures that enable us to think about the patterns of such relationships. In the 17th century, the invention of the calculus enabled Newton and his contemporaries to think about accelerated motion. Today, nonlinear dynamics provides a formal structure with which to see a kind of order in what appear to be disordered phenomena, such as changes in the weather. Theoretical work takes place in conversations and on office blackboards, and is executed in pen and paper or, increasingly, with a keyboard, computer and monitor. Philosophers of science disagree about whether theories reveal the fundamental structure of the universe or simply give us the tools with which to predict the future disposition of phenomena from their present behavior. Whichever of these views is correct, theories are not just products of inquiry, but essential elements in the laboratory, helping researchers connect the disparate results achieved by experimentation and suggesting new experiments.

Just as experiment needs theory, theory needs phenomena: theory (except for theories in pure mathematics) is about substances and processes whose properties must be ascertained by observation. Fieldwork, now practiced mostly by ethologists, ecologists and geologists, involves the attempt to observe the behavior of plants, animals and rocks in their ordinary or "natural" settings: the succession of trees and grasses in a meadow, the social behavior of baboons or gazelles, the history of a mountain range revealed in a highway cut. But fieldwork, even though conducted *in situ*, provides both too much and too little information. Entities in their natural settings are enmeshed in complex patterns of interaction. What we encounter in nature are those complex patterns. Even when we set out to observe the life-cycle of a single plant, what we observe is produced by the seed of the plant, to be sure, but also by the contents and structure of the soil, by the contents of the atmosphere and the temperature of the plant's surroundings. And these factors are

-86 Degree Freezers, 1995
12 panel typology
detail
(cat. no. 2)

Fossils, 1995
9 panel typology
detail
(cat. no. 3)

in their turn the outcomes of complex processes. Not only is there too much to describe, making some sort of selection and categorization necessary, but sorting out one process from another—the effects of excess heat from those of excess fertilizer, or those of air weight, say, from the effects of air density—is well nigh impossible when we're simply presented with the composite result of their joint action. Furthermore, if the Gaia hypothesis is correct, all these categories of process—mineralogical, biological, atmospheric—are interdependent aspects of the life-cycle of the planet.[1] The field is too full and our time in it too short.

The laboratory, then, is where natural phenomena, too big, too small, taking too much time or too little, are transformed into elements amenable to study. Objects and entities are isolated, broken down, recombined and subjected to stresses outside their normal range in order to accomplish a more precise delineation of their capacities. Even more, like the alchemists' workshops of an earlier time, the laboratory is a place for the recreation of rare phenomena, such as insulin and growth hormone, and the construction of new phenomena such as super-conducting materials and frost-resistant strawberries. Whereas the alchemist's dream of effecting the transformation of lead into gold was a chimera, contemporary chemists, molecular biologists and metallurgists, riding on centuries of accumulated experience, are transforming not just objects in the laboratory, but the conditions of contemporary industrial and post-industrial life.

Anyone can have an idea; any community can collectively embrace an idea. Hoping to make distinctions that matter, we reserve the term "scientific" for knowledge produced under certain kinds of constraints. Philosophers, historians and other scholars differ as to what these constraints are or should be and on the degree to which they do or should constrain. Some scholars focus on the differentiating aspects of scientific inquiry—features of inquiry that might guarantee the truth or objectivity of its outcome (standards of proof, canons of evidence) or, if not truth, at least reliability. Other scholars emphasize the commonalities of inquiry with other human activities: its immersion in culture, its vulnerability to faddishness, its relations with power. Both are right. Hypotheses must survive critical testing before being accepted as knowledge, and communities of inquiry are bound by their adherence to certain standards of inquiry—standards which determine what tests hypotheses must pass. Beyond such common criteria as observational and experimental evidence and avoidance of self-contradiction, however, there are no general standards that hold in all times and all places, standards that could be taken as definitive of science. While precision of measurement has latterly been stressed, the degree of precision required depends on the context of measurement and on the instruments available for its performance. We have come a long way from water clocks to atomic clocks, from inches to angstroms. On the other hand, excessive precision may so reveal the irregularities in regularities that we settle for less than is possible in order to include as many instances as possible in one model.

Simplicity is frequently invoked as a hallmark not just of scientificity, but, when two or more hypotheses or theories are in competition, of the truth or greater likelihood of truth of one in comparison with the others. But when we try to pin down what might be meant by "simplicity," it's clear that there are different notions and different measures: the order of equations, the number of different kinds of entities,

49

the number of different kinds of properties or of processes. A theory could be simple on one of these definitions but not on another. And, at times in the history of science, simplicity has been overridden in the name of other goals. Philosopher Nancy Cartwright warns against allowing theories about the ideal form of knowledge to dictate in advance the forms of nature.[2] How, after all, can we be sure that nature is simple rather than complex, austere rather than fecund? Explanatory power, by which is meant the capacity of a single hypothesis to encompass a variety of phenomena, is also advanced as a criterion for distinguishing a hypothesis worthy of acceptance from those less worthy. This facilitates what some claim is an overriding goal of scientific inquiry: unification. But should having one theory of everything be a goal? Why should we think that there will be one set of relationships that underlies all the variety of observable processes?

The gap between our evidential resources and explanatory aspirations persists and apparently formal criteria turn out to require the importation of other values or of substantive metaphysical assumptions. Inference from data to theory is mediated by background assumptions that reflect the beliefs and values of those engaged in inquiry, who themselves are members of societies, of cultures and of subcultures. What can prevent the deliberate or accidental dominance of an idiosyncratic set of assumptions is critical interaction among members of a scientific community and across different communities attempting to describe and explain the natural world. If this is so, the image of the lone genius providing insight into the secrets of the natural world is only partially correct. It takes genius to formulate some of the great theories of Western science: the physics of gravity, the theory of evolution by natural selection, the special theory of relativity. But what determines whether they count as scientific knowledge, whether they become the frameworks and foundations of theoretical syntheses and resources for future work, is the extent to which they survive the critical scrutiny of a community of inquirers and become thereby incorporated into a body of knowledge. In the cultural history of the West, there are anticipations of many of the notions central to contemporary science: Aristarchus, in the 3rd century bce, thought the sun the center of the planets' orbits, Democritus thought the world was constituted of atoms, Leibniz rejected absolute space and time. Without networks of practitioners, theories, relevant problems and techniques of verification with respect to which they might be elaborated and tested, these reasonings and speculations could not be accommodated into an ongoing systematic practice of inquiry. They would remain curiosities, singularities, rather than discoveries, which are credited to Nikolaus Copernicus, John Dalton, Marie Curie, Lise Meitner and Albert Einstein. What is identified as insight depends on a preexistent set of understandings, and the genius is the one identified by a community as providing it with a theory or model that simultaneously challenges and synthesizes those understandings. Knowledge is constructed by communities through interactions with the world and through the discursive, social interactions that constitute criticism, challenge and agreement.

Inquiry is thus inherently social, dependent on social interactions for the elaboration of networks of knowledge as well as for the assurance, as much as is possible, of reliability and objectivity. This is cognitive or epistemic interdependence. But there is another set of dependency relations between the sciences and the societies that support them. In earlier centuries science was the province of "gentlemen"

Sequential Molecules, 1995
9 panel typology
detail
(cat. no. 1)

who could afford the habit. In 20th-century industrial societies, science has not only fully entered the universities, but is primarily supported by the state and by industry.

Scientific research has become indispensable to advancing certain aims of states (defense and warfare), of corporations (the development of new materials, new substances and the means to mass-produce rare substances) and of societies (the control of disease). Just as these bodies have extended support for scientific research, so has research, especially experimental research, become dependent on that support. The Manhattan Project, which mobilized hundreds of scientists for a sustained effort lasting years, launched not just a new weapon, but a whole new way of doing science—replicated and varied in the fifty years since in physical and information research for the military, in the development of biotechnology at the intersection of industry and the academy, and now in the human genome project—a global effort to articulate the entire sequence of base pairs (the basic building blocks of the DNA molecule) of the human genome. In light of such multiple dependencies, how does who does science matter? While the new forms of support mean that its practitioners need no longer be independently wealthy, it must be noted that science is still largely the province of men, who have had the benefit of appropriate educations. Women are only slowly being integrated into the scientific workforce and most slowly at the top, where power and prestige reside. Members of traditionally underprivileged racial minorities are even more rare. Is it possible to disentangle the interests of the funders from those of practitioners when we try to understand how science is done? Whose interests determine what of the world will be revealed and what will remain unexplored? And who will benefit from and who will bear the burdens of these decisions?

In the 16th and 17th centuries, natural philosophy in Western Europe was shifting from giving an account of natural phenomena grounded in first principles to giving accounts of phenomena grounded in observation. Principle—theory—was and is, however, still important: the Royal Society in England dutifully made note of many observations of odd and singular phenomena that never made it into systematic accounts of nature. Without a theory, a model or a regularity—a run of similar instances—two-headed calves and other phenomena remained outside the reach of scientific understanding.

Experiment is a means of establishing regularities. These in turn require and permit the elaboration of more encompassing models and theories. Knowing nature's regularities is an essential step both for predicting their occurrence and for recreating them. In the laboratory, it is possible to hold many factors constant, to reduce the complexity of the natural site and observe the effect of one factor, one intervention. Substances can be passed through instruments, such as centrifuges or enzyme baths, which break them down into constituent parts, or that speed them up, as do particle accelerators. As we pursue the secrets of matter into ever smaller bits, new instruments and new technologies are required to help us read the results of reactions too small or too swift to be recorded by the unassisted eye. Gel electrophoresis passes an electric current through a sample of DNA, producing distinctive banding patterns that we take to reveal the particular sequence of the nucleotide bases—adenine, thymine, cytosine, guanine—constituting that particular strand. Detectors attached to particle accelerators sift through thousands of events to find a sought-for particle. The data of radio telescopes are transformed into images of a portion of the sky. But nature is not really laid bare by these machines. At best they present us with the traces, the signatures, left by nature. The machines themselves are theory-laden windows onto the world and they require us to use our theories to reconstruct the causal chain from particle or gene structure to the images that we take to be their traces. Without the theory, we have only unintelligible strings of marks, or smudges. What I see when I look at a bone marrow smear is very different from what a laboratory technician who notices differences from one smear to another sees, which is in turn very different from what the cancer researcher or the pathologist who interprets those differences sees.

In some cases the instruments represent short cuts—ways to consign hours of painstaking and tedious labor to automation, like the DNA sequencer, which, using lasers and fluorescent labelling, can read thousands of base pairs (adenine-thymine; cytosine-guanine) of DNA in a day.[3] In others they represent theoretical projections into the world, such as the particle detectors or radiotelescopes which transmit what our theories say is evidence of pions or top quarks or of quasars and black holes. We think of scientific technology as hardware—glass, metal and ceramic— but in molecular biology, researchers have learned to harness biological processes themselves in service of research. As just one example, the polymerase chain reaction enables researchers to reproduce desired segments of DNA with extraordinary efficiency—up to a million copies of a given segment in twenty cycles lasting four to five minutes each. Previously, desired quantities of DNA could be produced only by taking advantage of the reproduction cycles of bacteria or yeast (no mean biotechnological feat in itself). Once these processes are developed they seem obvious. Their rapid incorporation into the laboratory leaves no trace of the struggle to develop and perfect them—the hours spent with test tubes and reactants, the false starts and promising, but finally less efficient, less accurate processes that accumulate like used drosophila in the wake of any successful scientific development. The successes—monoclonal antibodies, oncomouse, yeast artificial chromosomes (known as YACs)—become mundane aspects of a research process which rolls on to newer challenges and newer accomplishments.

In this process, certain parts of the natural world, because they can be stabilized and reproduced, come to stand for the whole. The construction of knowledge proceeds across networks of laboratories, some in competition, some in collaboration with each other. Assurance that they are addressing the same phenomenon comes in part from the development of standardized research tools and products: genetically identical fruit flies (the heroes of 20th-century genetics), mice or bacteria that will respond in the same way to identical interventions whether performed in San Francisco, St. Louis, Paris, Buenos Aires, Tsukuba or Zurich. These standardized tools, maintained by heroic measures, while themselves the products of a fragmentation of nature, are what permit the transcendence of locality and the integration of particular experiments performed in particular places into global, international science.

But if the objects about which scientific knowledge is constructed are thus isolated from their ordinary environments and interactions, preserved in their isolation by measures like freezing or moisture and atmosphere control or immersion in formaldehyde, and subjected to strains and stresses intensified beyond the range of their "natural" state, to what extent can we think of science as accounting for

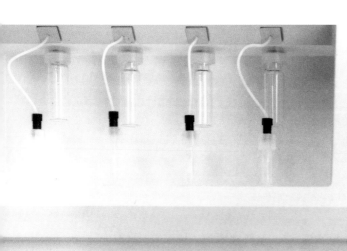

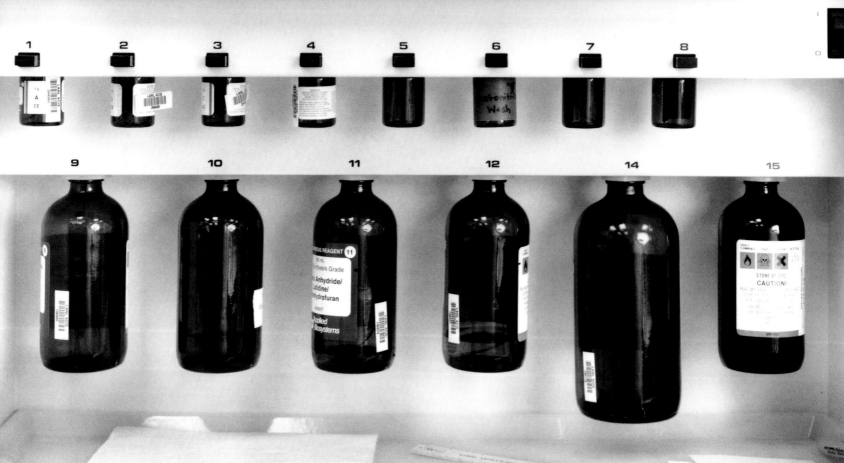

nature as it is in itself, as contrasted with nature made malleable in the laboratory? Of course, field studies and clinical trials act as a check on our generalizations beyond the laboratory, but with philosopher Joseph Rouse we might ask to what extent we transform our own environments to accommodate the objects and the products of laboratory knowledge.[4] Is the sterilized space of the glove box just an extreme example of the construction of safe environments for the new tools of our trades?

Bruno Latour and Steven Woolgar, posing as naive observers, asserted, partly tongue in cheek, that the function of a laboratory was to produce inscriptions.[5] For example, in an endocrinology lab, tissue from many animals' thymus glands, or pancreas or pineal glands, is ground up, possibly treated with fluorescence or a radioactive marker, and put into a vessel for heating or shaking or cooling or interacting with a better known substance. The machine is connected to a stylus or other recording device that generates marks, inscriptions on a surface of paper, metal or silicone. The laboratory members discuss the marks, compare different such outputs, construct a composite set of marks and insert them into text, which is then shipped out of the laboratory. The laboratory, in turn, receives similar graph- and table-speckled texts from elsewhere. We recognize in this description the production of a research paper, the medium for communicating scientific knowledge, and think we share the joke.

But if the function of the laboratory is not to process, produce and circulate inscriptions, if the above is too thin an account of all that activity, what is it? Ask an individual researcher and you will get a variety of answers. A young scientist I knew in graduate school was quite candid about his aims: he wanted to receive the Nobel Prize. James Watson, perhaps my friend's model, was quite clear about the glory that would attend whoever cracked the puzzle of the structure of DNA.[6] But other scientists will talk of their desire to know how things work, of their love of nature or of a compelling need to understand the universe, an overwhelming curiosity about the makeup of things or of the sheer pleasure of having a recalcitrant puzzle give way to persistence and insight and daring. Those more psychoanalytically oriented may acknowledge a need for power and control that is satisfied by the manipulation of substances and instruments, by precise measurement. These individual motives tell us why particular individuals choose to pursue careers in science, but cannot explain why the sciences have come to dominate intellectual life in industrial societies. Certainly the power of the sciences to illuminate the contours of vast quantities of phenomena, a power granted by the union of formal mathematical structures with careful and controlled observation, is a part of the source of this cognitive authority. Something else, however, must explain the social and economic preeminence of science—the popular interest that sustains numerous publications, from *Scientific American* to *Discover* to every Tuesday's "Science Times" in *The New York Times*, and television shows like *Nova*, and the power to command the lion's share of academic resources in universities and colleges. Members of Congress might want to reshape the National Science Foundation and National Institutes of Health, but they wouldn't dream of closing them down as they have threatened to do with the National Endowments for the Humanities and for the Arts.

Just over fifty years ago, "the Bomb" jolted the common perception of scientific research. No longer purely a rarefied, ivory tower pursuit, it became an activity

DNA/RNA Synthesizer, 1992 (cat. no. 27)

whose effects could permeate every household, from the fear of atomic attacks to the harnessing of previously unimagined forces to human bidding. The jury is still out on the advisability of having proceeded with the development of nuclear power, but there is no doubt that both as a material force engaging supporters and resisters, and as a focus of a modern imagination entranced by concepts of relativity and indeterminacy, the outcomes of physics research of the first half of this century indelibly reshaped the world. In the second half, developments of wartime information and operations research quietly and gradually facilitated the present infiltration of almost every facet of industrial and intellectual life by computers. Subatomic physics, chemistry and information science have also transformed scientific research, making possible the acquisition, processing and transfer of data in quantities and speeds that increase yearly. In the last third of this century they have been joined by biology, especially molecular biology, as engines of change.

Because it reaches deeply within the living organism, because frequently pioneering research and technological development are simultaneously achieved in the same project, because the pace of change has been so rapid, the new biology poses challenges that seem even greater than those presented by the atomic and information revolutions. As a symptom of the seriousness of this challenge, molecular biologists themselves adopted a short-lived moratorium on recombinant DNA research in the mid-1980s. While the moratorium itself did not last, its legacy is national guidelines and (usually quite modest) local controls on the uses of genetic engineering. In the meantime, we are promised frost-resistant strawberries, tomatoes that will ripen off the vine and the quadrupling of a dairy cow's milk-productive capacity. These developments and the inevitable others are just the first installments. The new biology has learned to use natural processes at the micro-level, the subcellular and genetic level, to free us from the old constraints of nature at the macro-level, not just the inevitability of the seasons or the vagaries of the weather, but also the costs of industry. Strains of bacteria are proving capable of consuming dangerous wastes, their metabolic systems breaking down toxic oils and pollutants into harmless by-products. In addition to thus transforming various facets of production, molecular biology is producing new understandings of human health, disease and identity. It is these which merit not just notice, but a level and quality of reflection in which we may be ill-prepared to engage.

The possibilities of genetic manipulation have given a point to the search for the genetic basis of various human afflictions. Thus we hear of searches for the gene for breast cancer, for colon cancer, for alcoholism, for manic depression, even for (homo)sexuality. But genes code for proteins and in only a few cases is the presence or absence, or the over- or underproduction, of a single protein the cause of a given disease. And even when an inherited mutation in a single gene is the isolatable cause of a disease, it may, as in breast cancer, be associated with a small percentage of all the cases of that disease.[7] Much more common are diseases that involve multiple genes, both structural and regulatory. Furthermore, even though there are some mutations that will inevitably produce disease, in most cases the organism's own physiology and external environmental factors will modulate its expression from barely noticeable to cripplingly severe. The biochemical or molecular genetics of a disease may well provide the basis for a cure or preventive therapy. In all but a few cases, such therapies lie tantalizingly in the future. In the meantime, what is possible is termi-

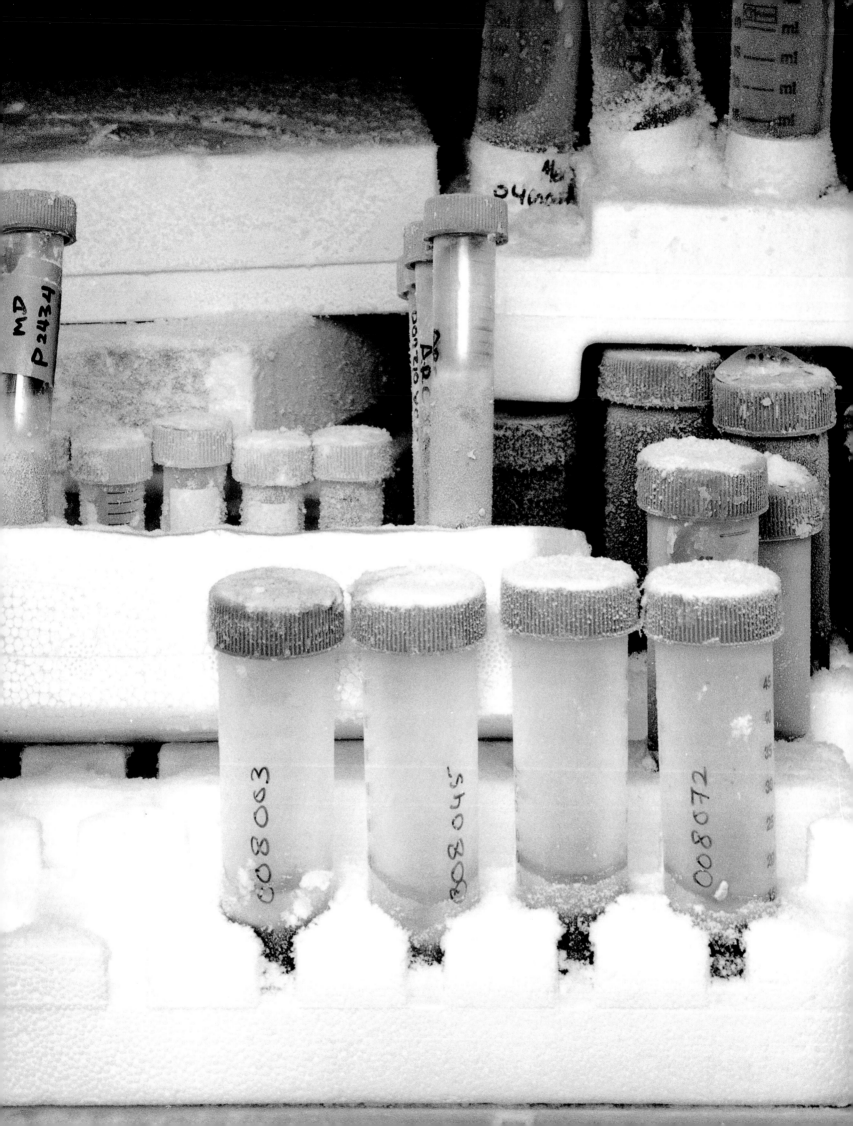

nating the gestation of a fetus identified as genetically predisposed (with some degree of probability) to some given adverse condition. This seems, to some at least, a blessing if it means less suffering from fatal and painful diseases. But it is a blessing at best mixed. The present ability to identify the sex chromosomes of a fetus is resulting in an underrepresentation of females in some societies. While most physicians now recoil in horror from past attempts to control unwanted behavior by performing hysterectomies and lobotomies, it is not clear whether that past will serve as a lesson or as a precedent.

-86 Degree Freezers, 1995
12 panel typology
detail
(cat. no. 2)

The human genome project is partly modeled on the Manhattan Project—a huge dispersed undertaking involving large numbers of researchers and many facets of research. It is always described using big numbers: three billion dollars (the amount appropriated by the United States government), three billion base pairs, one hundred thousand genes. Its initial aim is to characterize the human genome by specifying the complete sequence of nucleotide bases that constitute the genes. As was the case in the Manhattan Project, spin-off benefits are expected in the course of pursuing the central aim. The sequence itself tells us nothing. Eventually, with the aid of the sequence, a map can be developed which would specify the functions of the segments of sequence that constitute genes. The project has already enabled the invention of significant new technologies for genetic research, and because of the need for animal models will also increase our knowledge of the genetic structure of other species. Just recently, the first delineation of an organism's entire genome was announced. The organism is the bacterium *Haemophilus influenza* with 1,830,137 base pairs constituting an estimated 1,743 genes. We have a long way to go before achieving the goal of sequencing the entire human genome. Participants in the project nevertheless hold out the vision of each of us eventually possessing a compact disk that contains the representation of our own individual genetic profile.

In the meantime, a portion of the funds appropriated for the project have been set aside to support research on the ethical and legal implications of the project. Some issues are clearly ethical and have already been raised, if not solved. Can genetic information be used to screen medical insurance applicants? Can an employer require the genetic signature of a prospective employee as part of her or his application for employment? These questions are often addressed under the rubric of privacy rights, which seem a thin and narrow focus on these questions. Should we not rethink the whole framework of risk that underlies the concept of medical insurance? Whose responsibility does health become? What would be the fate of those rejected for employment on the basis of presumed future disability? We already know that there is only a probabilistic connection between a given genetic profile and the manifestation of a phenotypic trait. Does not the kind of knowledge we think we may attain mandate a deeper investigation than one that holds everything constant except that little compact disk? Furthermore, if genetic information can only really give us a probability of a certain trait's appearance, just what is the privacy whose violation concerns us here? A family history, which, in principle, anyone knowing biological parentage could assemble, may be just as informative.

Other questions one might raise seem neither purely ethical nor purely scientific. Take, for example, the central aim of characterizing the entire human genome.

If everyone's genome is unique, whose genome will be characterized? A man's? A woman's? Of what ethnic or racial ancestry? And once these issues are addressed, there is the matter of how to think about the sequence produced. It is supposed to represent a typical, "normal" member of the species. But how do we determine what is normal? Are women who develop breast cancer or men who develop prostate cancer at age 75 normal or abnormal? Is a woman who develops osteoporosis in her 60's normal or abnormal? Is someone with a genetic profile associated with a 25% probability of developing cancer but who lives a long life and dies of heart disease normal or abnormal? Since characterizing the sequence is the first and necessary stage in producing a map, part of the point is lost if we stop to produce multiple alternative sequences. In any case, how much variation could we tolerate within our compass of normality? Is there anyone who doesn't carry a genetic predisposition for some disease state? Is normality determined by reference to genotype (the predisposition) or to phenotype (its realization)? That first sequence will be a composite, but a composite is a construct, not a naturally occurring object. As Evelyn Fox Keller has asked, who will decide what gets to represent the human standard?[8] And in what contexts is such a notion even meaningful?

Some proponents of the human genome project extol it as the ultimate in species and individual self-knowledge, and thus worth pursuing independently of any more material benefits. For them, the ethical problem is how to avoid the simplistic self-exculpation, "my genes made me do it," which they see as licensing all kinds of anti-social behavior. But this is far too deterministic for most philosophers, as for some biologists, and it is hard to see what a gene sequence has to do with the particular propaganda-driven barbarities of recent warfare, or with the emotional significance of a Brahms sonata, a Neruda poem or a DeFeo painting, which exists in the interaction of artist, artifact and listener, reader, viewer. It's worth comparing this view of self-knowledge with that of the 17th-century Dutch philosopher Benedict Spinoza.[9] Spinoza, too, was a determinist, but held that self-knowledge was the key to human freedom. Humans, Spinoza thought, are moved by desires for what we believe to be our good. But we are often mistaken as to what is really good for us and do not understand the nature of our desires. Human bondage consists in our being induced to action by causes of whose true nature we were ignorant. Once our ideas became adequate, that is, once we came to know our true natures (which, for Spinoza, involves knowing the true nature of everything), we would know not only what the causes of our (previous) desires were, but would no longer be their victim. Our actions would be determined, but also free, since determined by causes which we could, in our complete understanding, endorse as conducive to our genuine well-being. Spinoza's vision expresses a faith in reason largely abandoned in the 20th century, but its aspiration to transformative self-knowledge from the inside is both profoundly human and beyond the reach of any laboratory.

Science and ideas of science have profoundly shaped life in modern industrial societies, just as these societies have provided the material and cultural conditions in which a certain kind of science has flourished. Scholars debate whether the sciences that have developed in Western Europe and North America represent the epitome of natural knowledge or forms of knowledge inflected by the particular preoccupations of the West. As Western laboratory science becomes practiced across the globe—in Japan, India, China, Ghana, Kenya, Brazil, Argentina and elsewhere—this debate certainly grows more complicated. In part, it revolves around what kinds of representational and interventive practices get to be called "science," and how much freedom there may be to alter and redirect present practices. To a great extent this is an issue determined in the doing rather than in the abstract. But as the sciences and the products of science-based technologies reach ever more deeply into our lives, thinking people are asking how laboratory research and its outcomes are both enlarging and restricting the scope of human experience and aspiration. Catherine Wagner's photographs invite us to reflect not just about the effects of scientific research on us, but about the ways in which the sciences express the common culture we all participate in making.* ❖

*I wish to thank Carl Chung, Helen Donis-Keller, Valerie Miner and Elizabeth Spelman for helpful comments on earlier drafts.

NOTES

1. For more on the Gaia hypothesis, see James E. Lovelock and Lynn Margulis, "Atmospheric Homeostasis for and by the Biosphere: The Gaia Hypothesis," *Tellus* 26 (1974): 1-10. The Gaia hypothesis is presented in more popular form by James E. Lovelock in his *Gaia: A New Look at Life on Earth* (New York and Oxford: Oxford University Press, 1979).

2. Nancy Cartwright, *How the Laws of Physics Lie* (New York and Oxford: Oxford University Press, 1983). Other recent philosophical studies of science include Ronald Giere, *Explaining Science* (Chicago: University of Chicago Press, 1988); Ian Hacking, *Representing and Intervening* (Cambridge: Cambridge University Press, 1983); and Bas van Fraassen, *The Scientific Image* (Oxford: Oxford University Press, 1980).

3. For more on the automation of tedious processes in human genome research, see Leroy Hood, "Biology and Medicine in the Twenty-First Century," in Daniel J. Kevles and Leroy Hood, eds., *The Code of Codes: Scientific and Social Issues in the Human Genome Project* (Cambridge: Harvard University Press, 1992), pp. 136-63.

4. Joseph Rouse, *Knowledge and Power* (Ithaca, NY: Cornell University Press, 1989).

5. Bruno Latour and Steven Woolgar, *Laboratory Life*, 2nd ed. (Princeton: Princeton University Press, 1986). For other recent anthropological and sociological studies of laboratory science, see Andrew Pickering, *Constructing Quarks* (Chicago: University of Chicago Press, 1984), and Sharon Traweek, *Beamtimes and Lifetimes* (Cambridge: Harvard University Press, 1987).

6. James Watson, *The Double Helix* (New York: Atheneum, 1968).

7. The situation is more complicated than any quick statement suggests. Two genes (BRCA-1 and BRCA-2) have been identified that each are involved in about half of the cases of familial, or inherited, breast cancer, which constitute perhaps ten percent of all cases of breast cancer. Since their implication in inherited breast cancer, BRCA-1 and BRCA-2 have been intensely studied. They are thought to be involved in other cancers such as ovarian and prostate cancers; researchers have linked malfunctions in the protein produced by otherwise normal BRCA-1 with sporadic, i.e., non-familial, breast cancers; and one study has linked mutations in BRCA-2 with about 65 of a sample of 200 sporadic breast tumors.

8. Evelyn Fox Keller, "Nature, Nurture and the Human Genome Project," in Kevles and Hood, *The Code of Codes*, pp. 281-99.

9. Benedict Spinoza, *Ethics*, 1678, trans. W. H. White and A. H. Stirling (Oxford: Oxford University Press, 1927).

FOLLOWING PAGE:
-86 Degree Freezers, 1995
12 panel typology
detail
(cat. no. 2)

Plates

AA1

AA3

AA2

AG

Sequential Molecules, 1995, 9 panel typology (cat. no. 1)

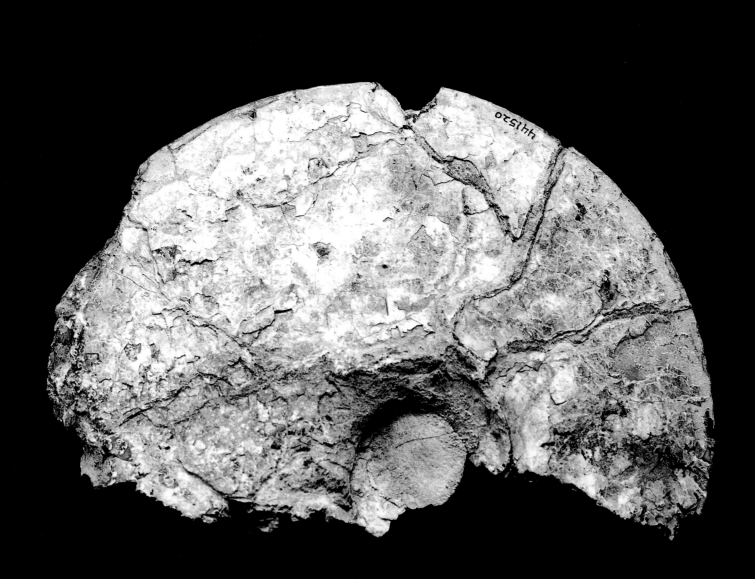

Body Pattern Formation, 1995
(cat. no. 13)

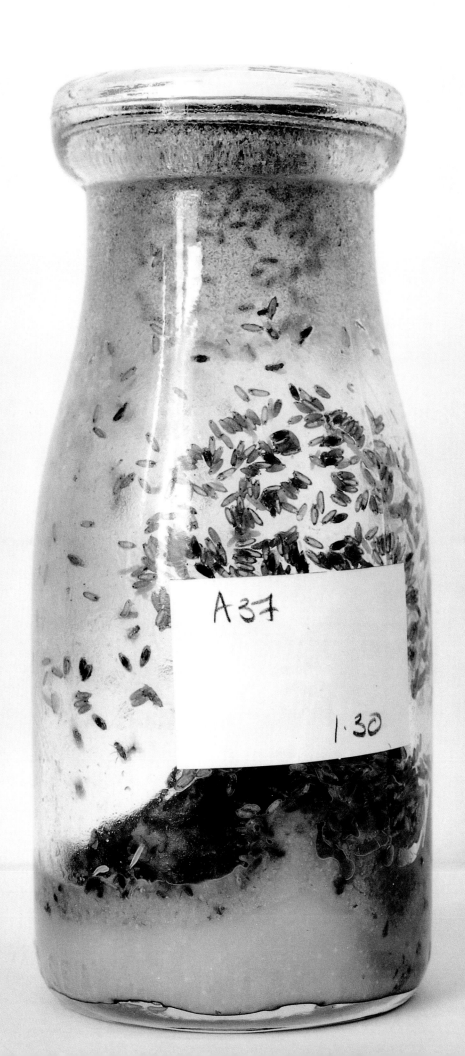

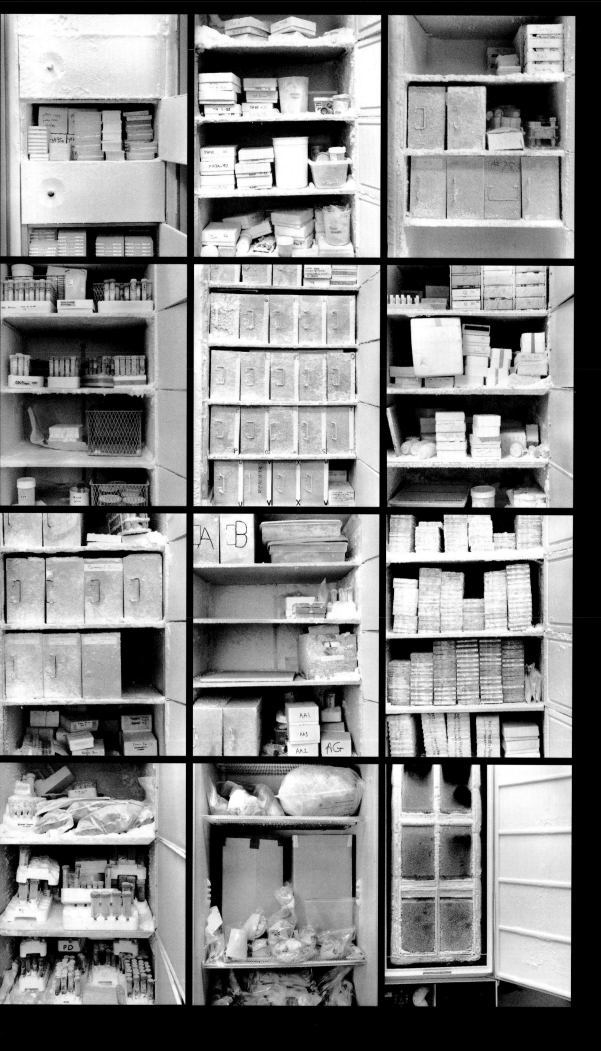

DNA CLone Library

Colon Cancer

Human Genome Project

Bipolar Disorder

HIV

Breast Cancer

Thyroid Cancer

Alzheimer's

Mega-Yac Library

Alcoholism

Tissue

-70° Freezer

-86 *Degree Freezers* (12 areas of concern and crisis), 1995, 12 panel typology (cat. no. 2)

Personal Boxes

Clone Archives
set A

Clone Archives
set A

Clone Archives
set A

Tumor Tissue
Box 4

Tumor Box
15

Tumor Tissue
#3

Tumor Box 13

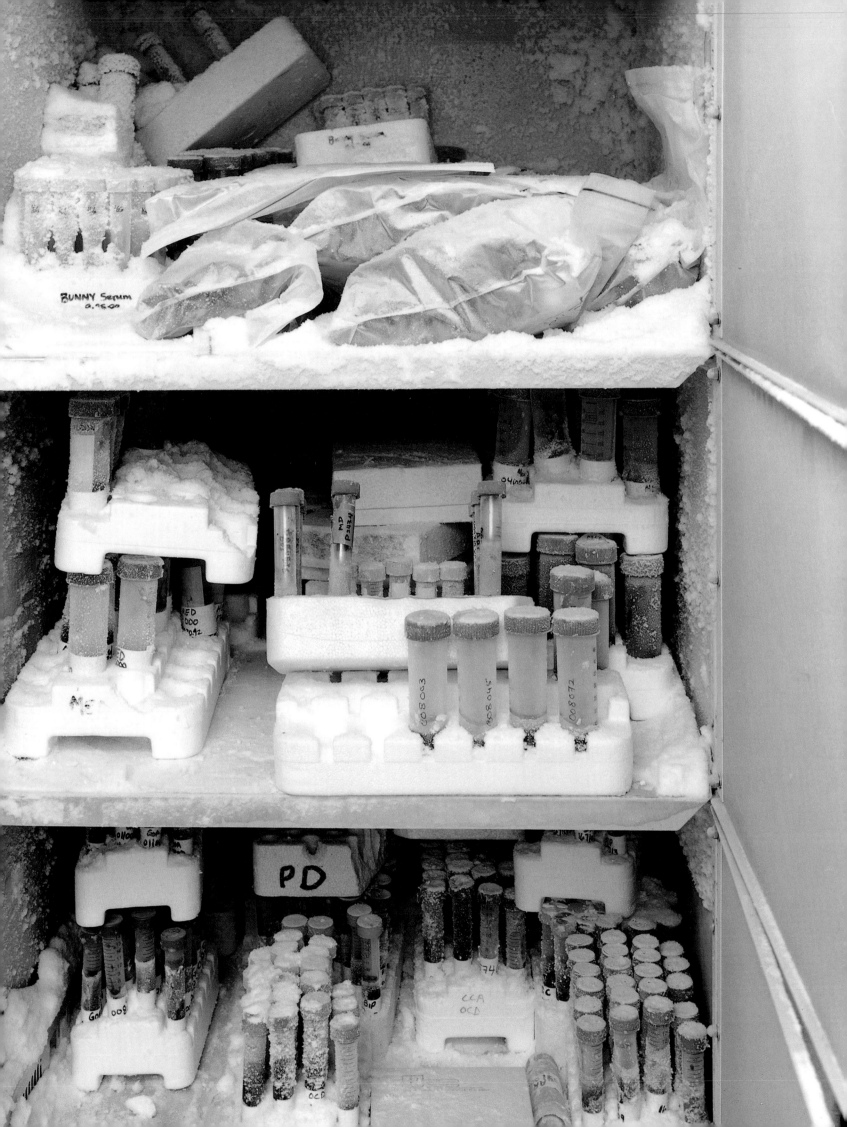

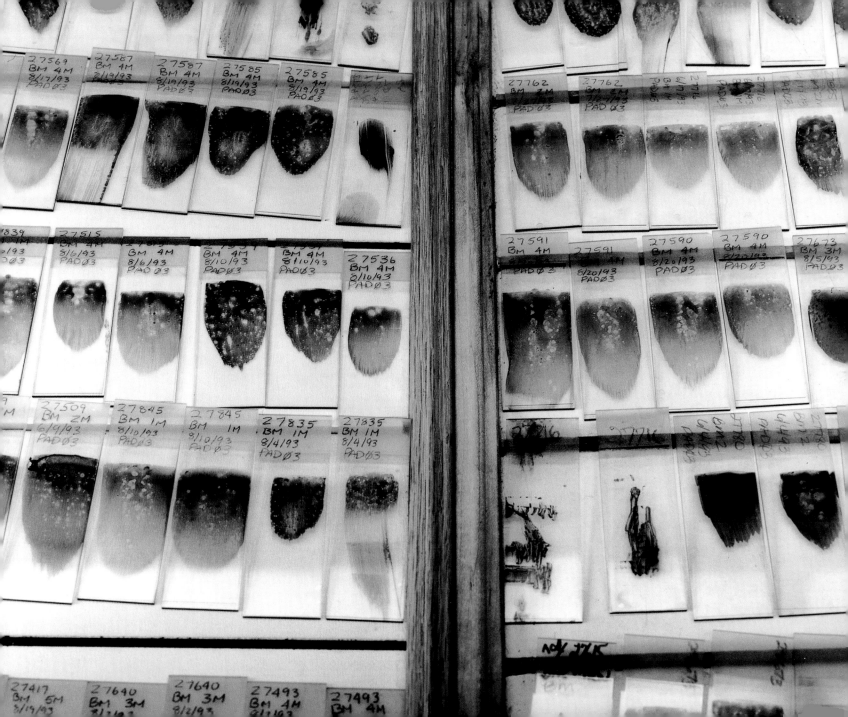

OPPOSITE:
Radioactive Cell Growth, 1994
(cat. no. 23)

FOLLOWING PAGES:
Beating Heart—Heart Chamber, 1994
2 panel diptych
(cat. no. 7)

90

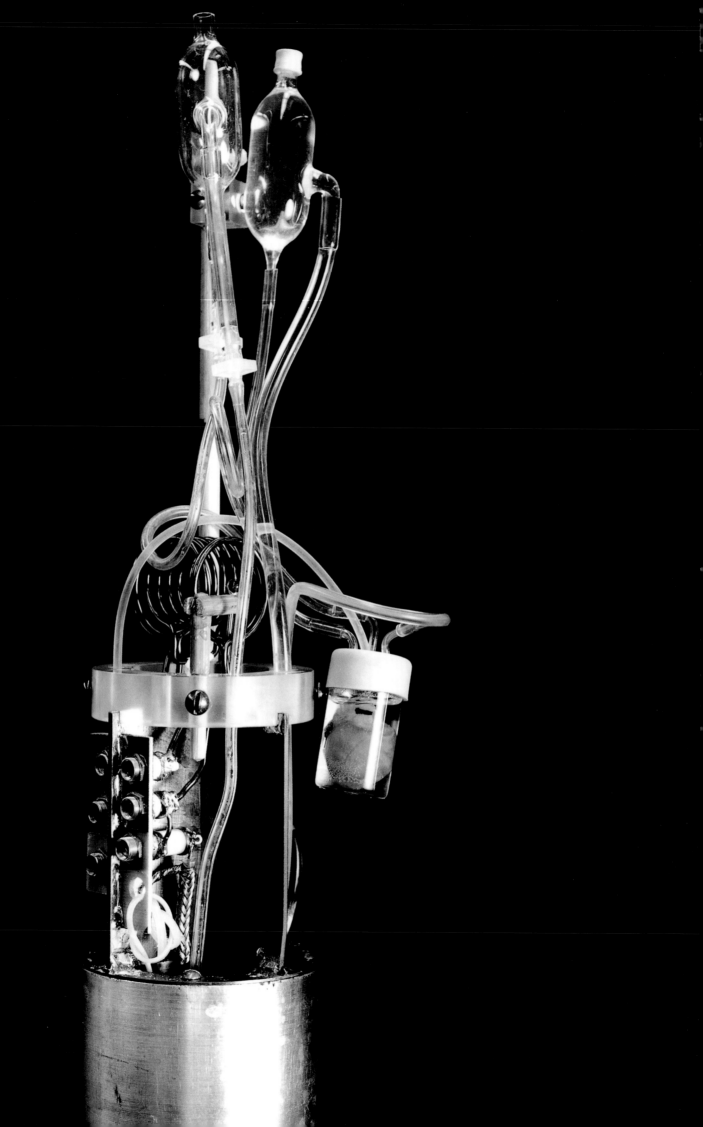

OPPOSITE:
Sex Linkage, 1995
(cat. no. 12)

FOLLOWING PAGE:
Definitely Not Sterile, 1995
(cat. no. 8)

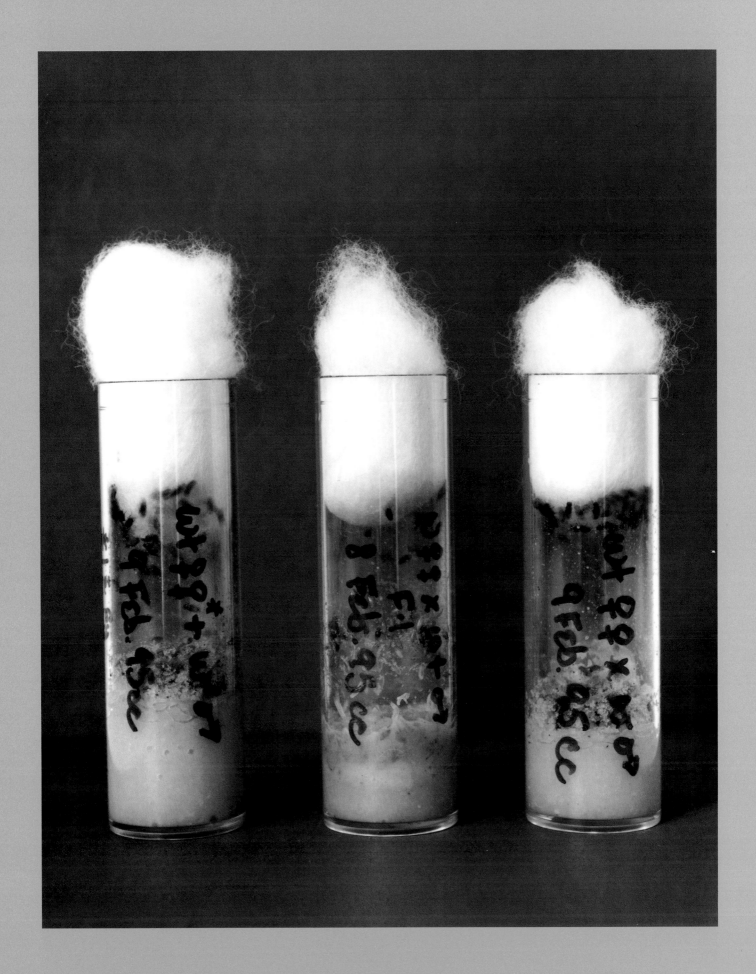

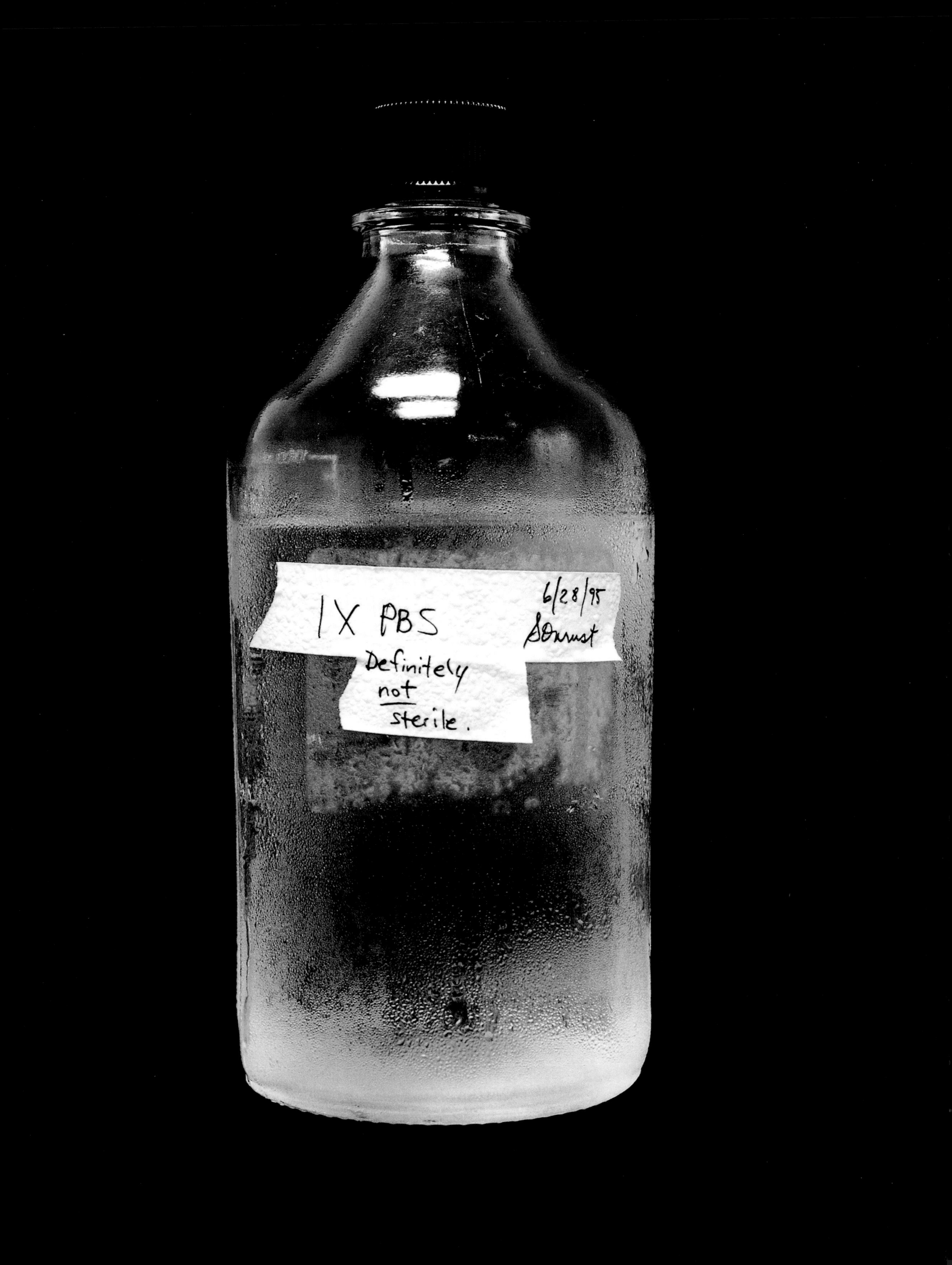
1X PBS
6/28/95
Definitely
not
sterile.

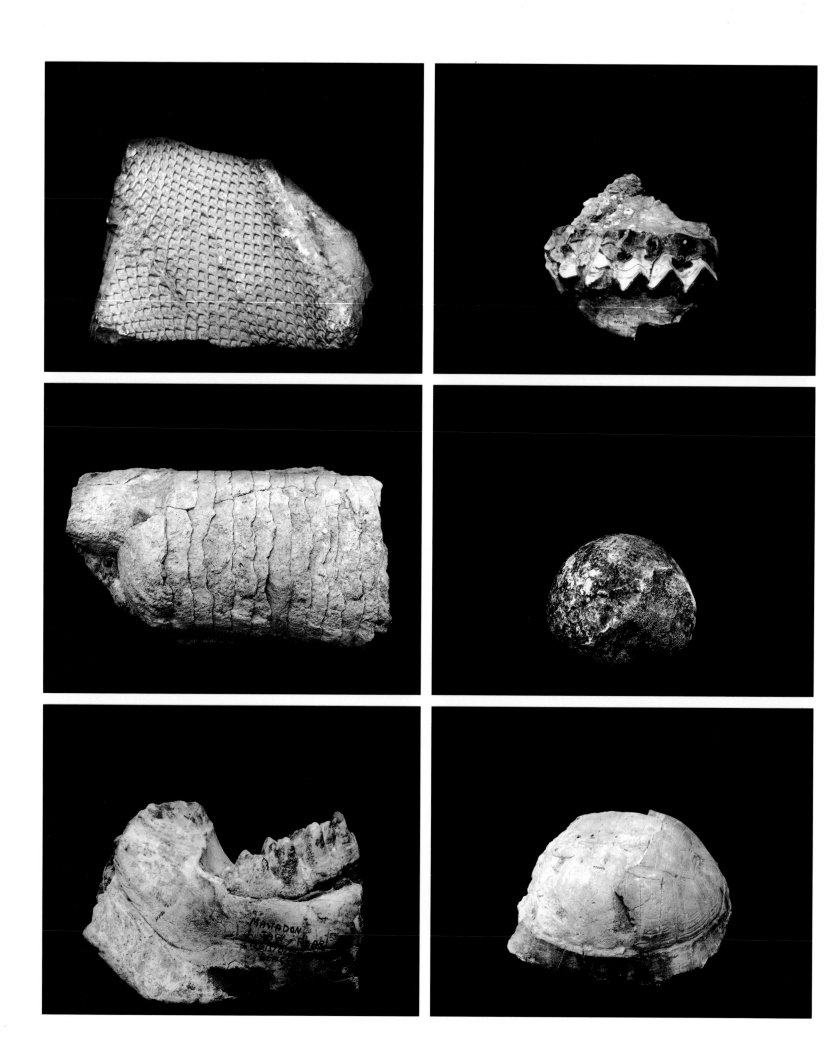

Fossils, 1995, 9 panel typology (cat. no. 3)

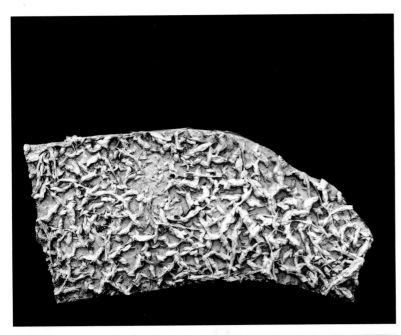

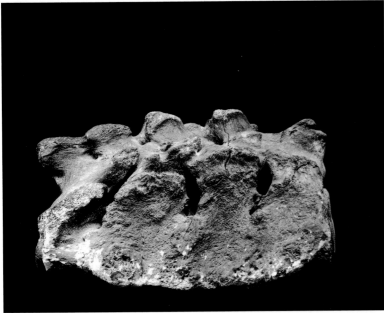

Scale Tree

Oredont Jaw

Crystal Casts

Cephalopod

Femur Fragment

Whale vertebrae

Mastadon Jaw

Turtle

Brontotherium Foot

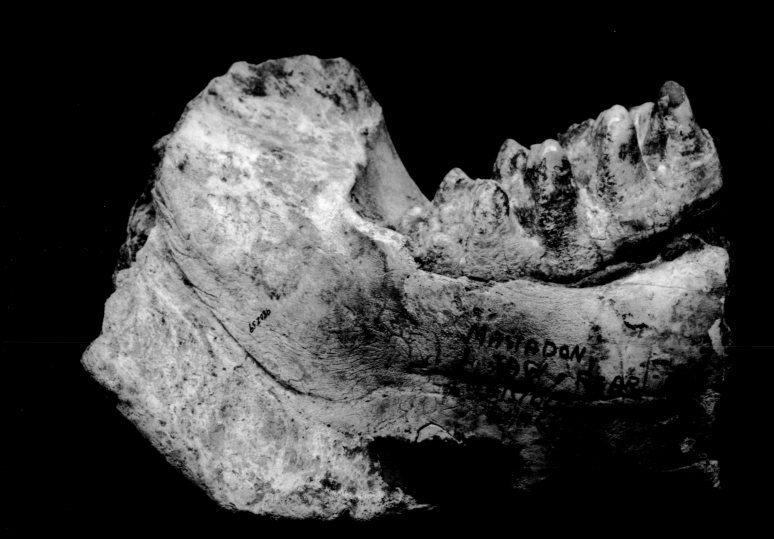

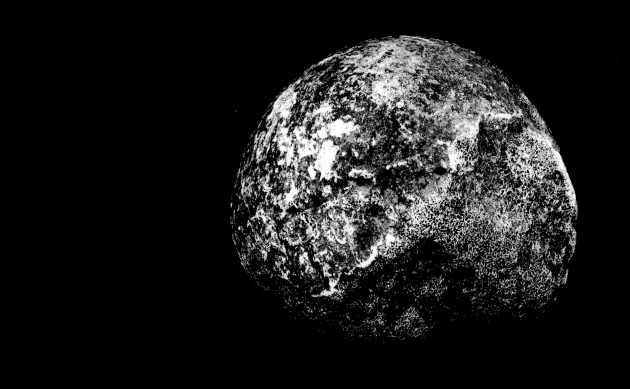

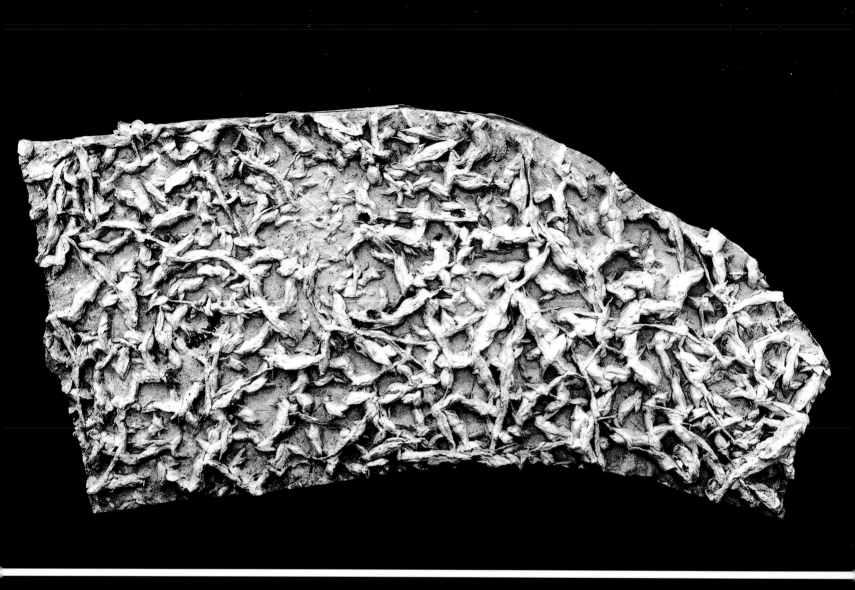

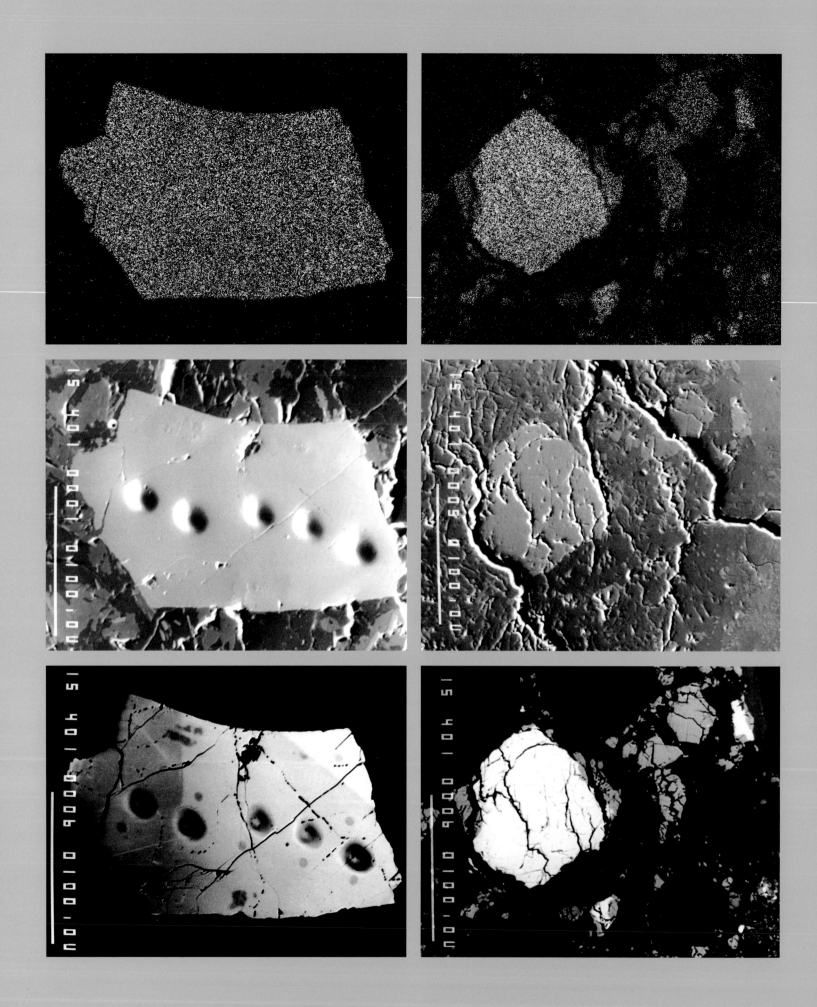

Moon Rock Found on Moon, Moon Rock Found on Earth, 1994, 6 panel typology (cat. no. 4)
Top: X-ray; Middle: Topographic analysis; Bottom: Chemical analysis

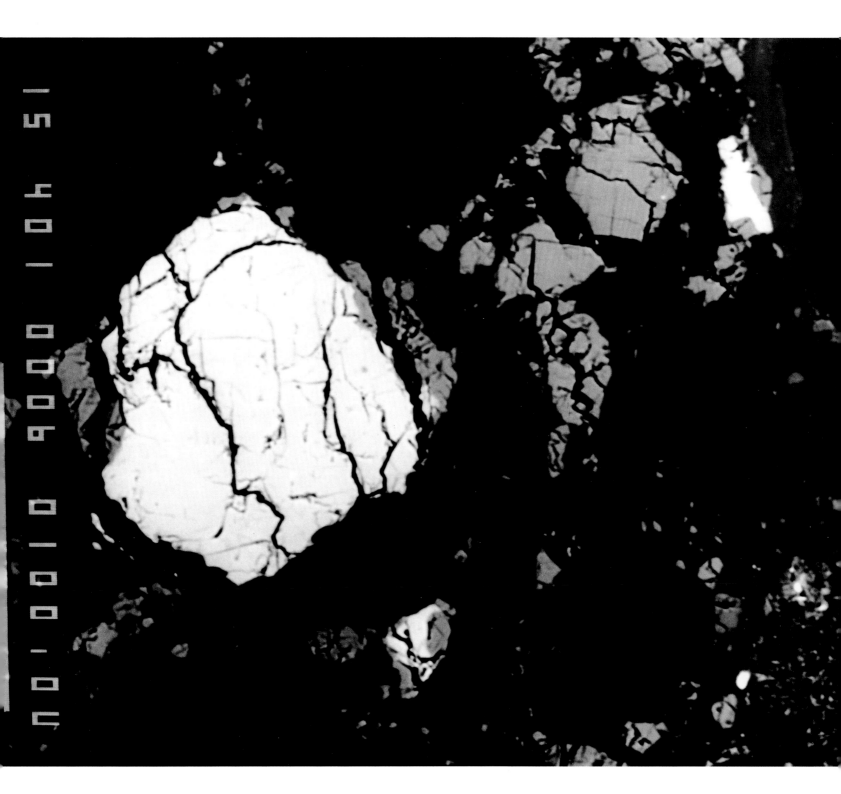

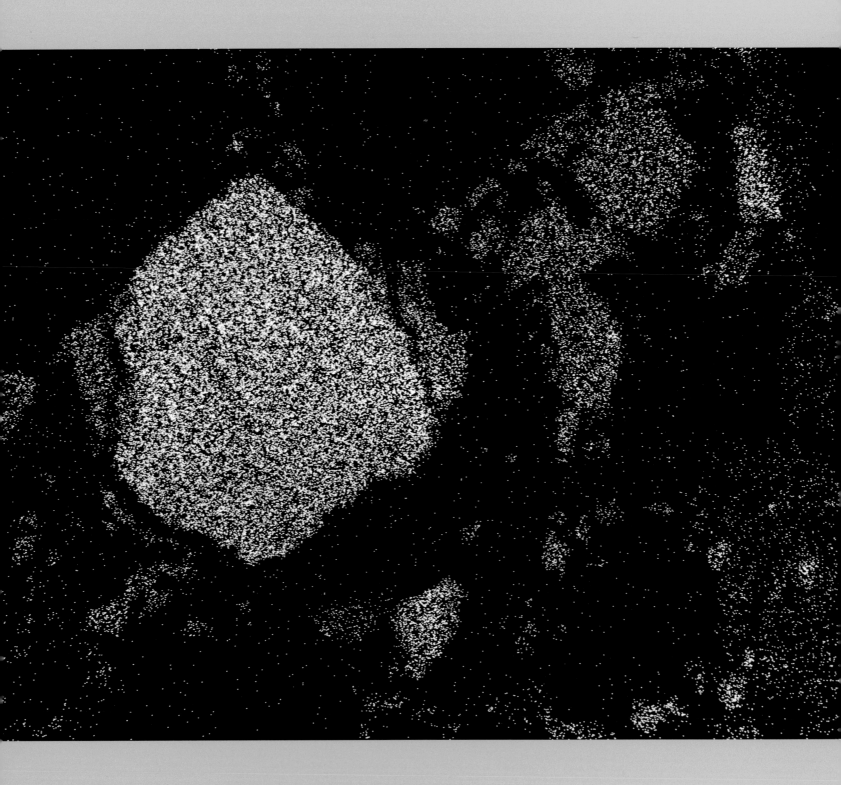

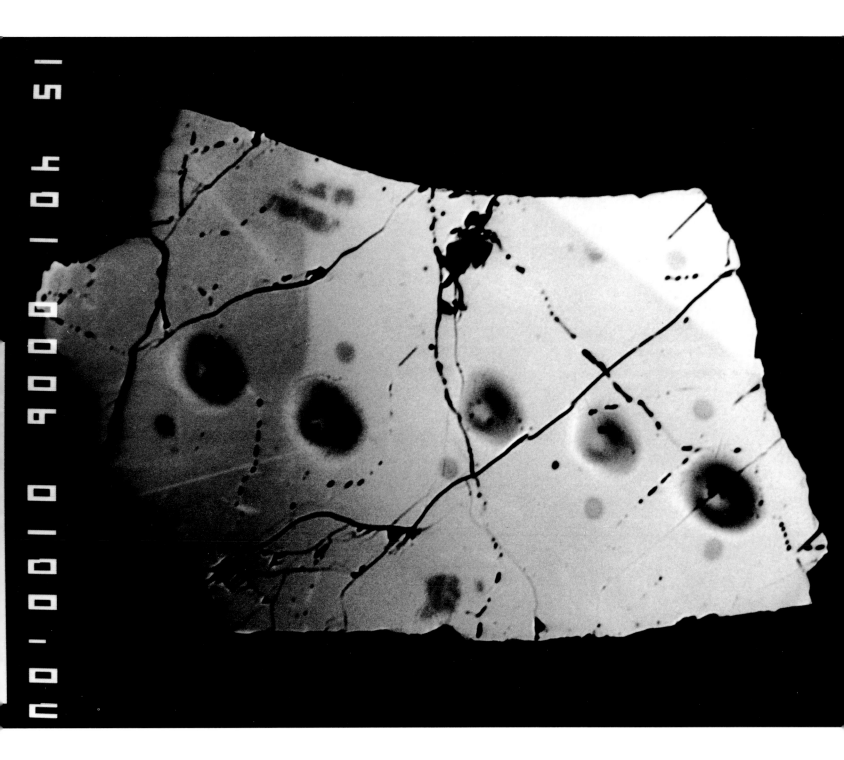

107

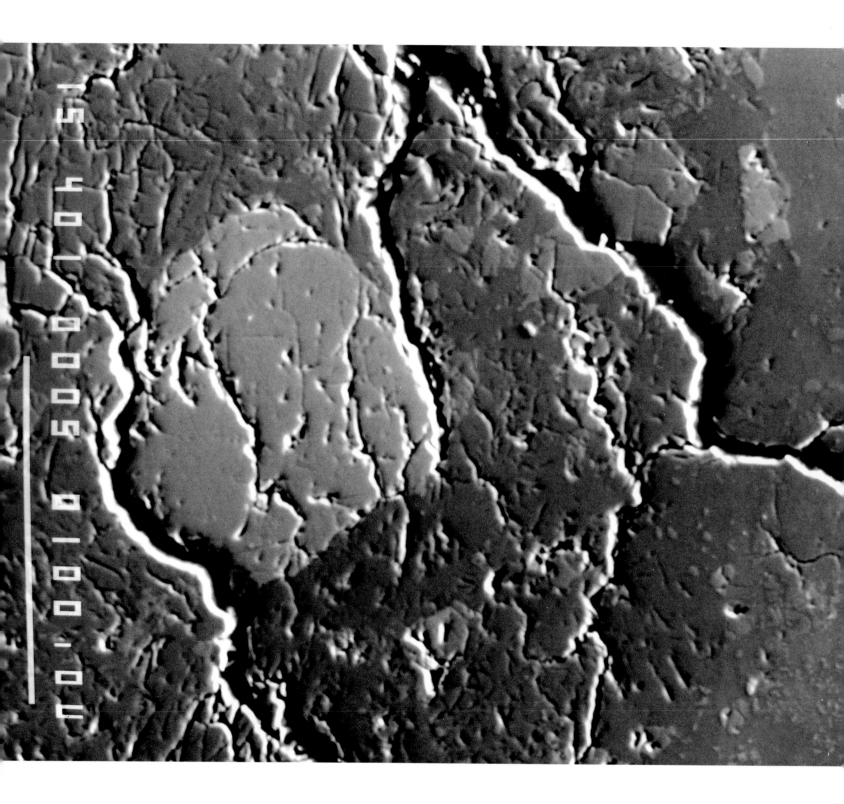

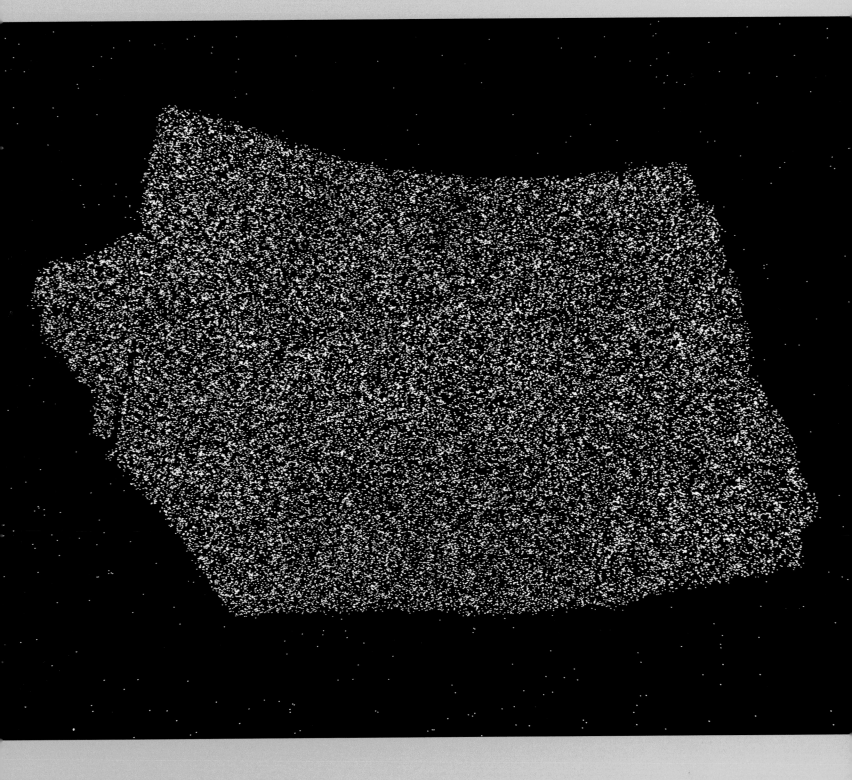

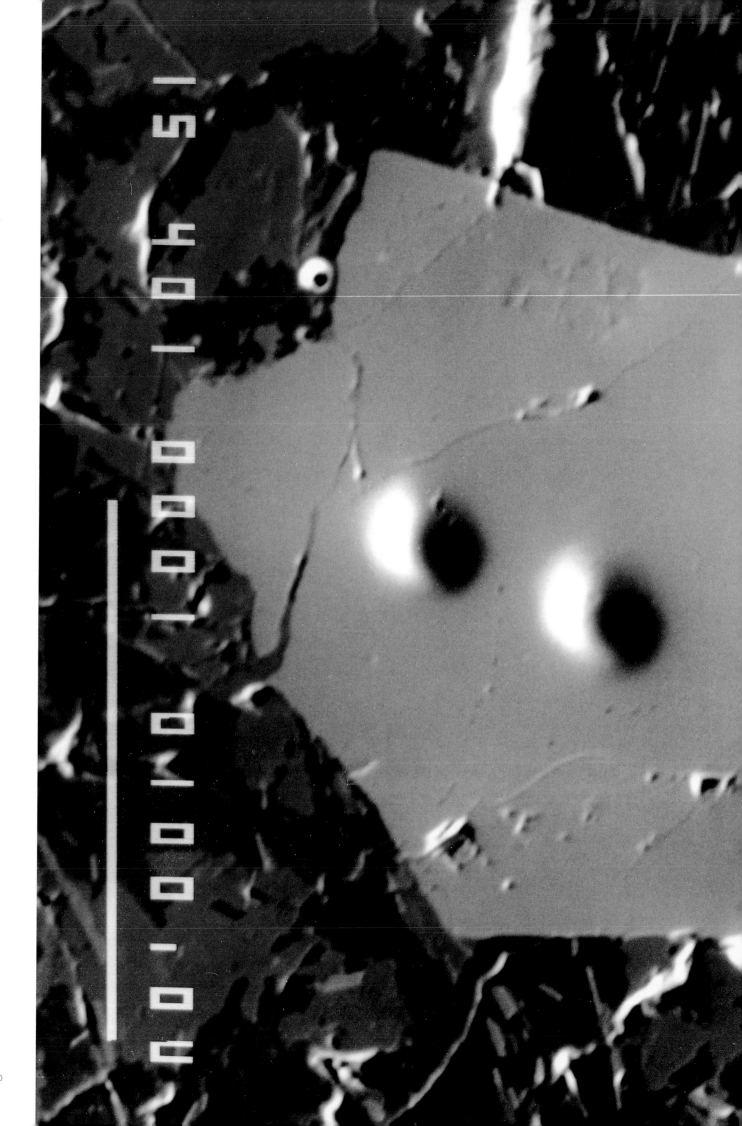

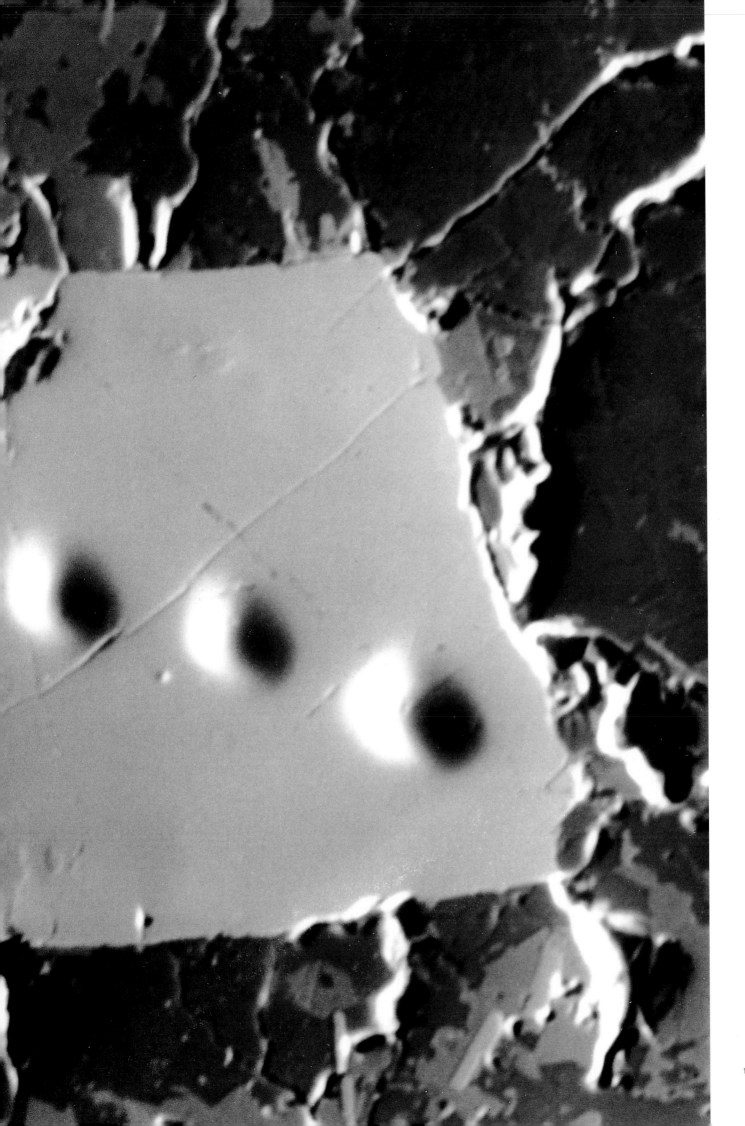

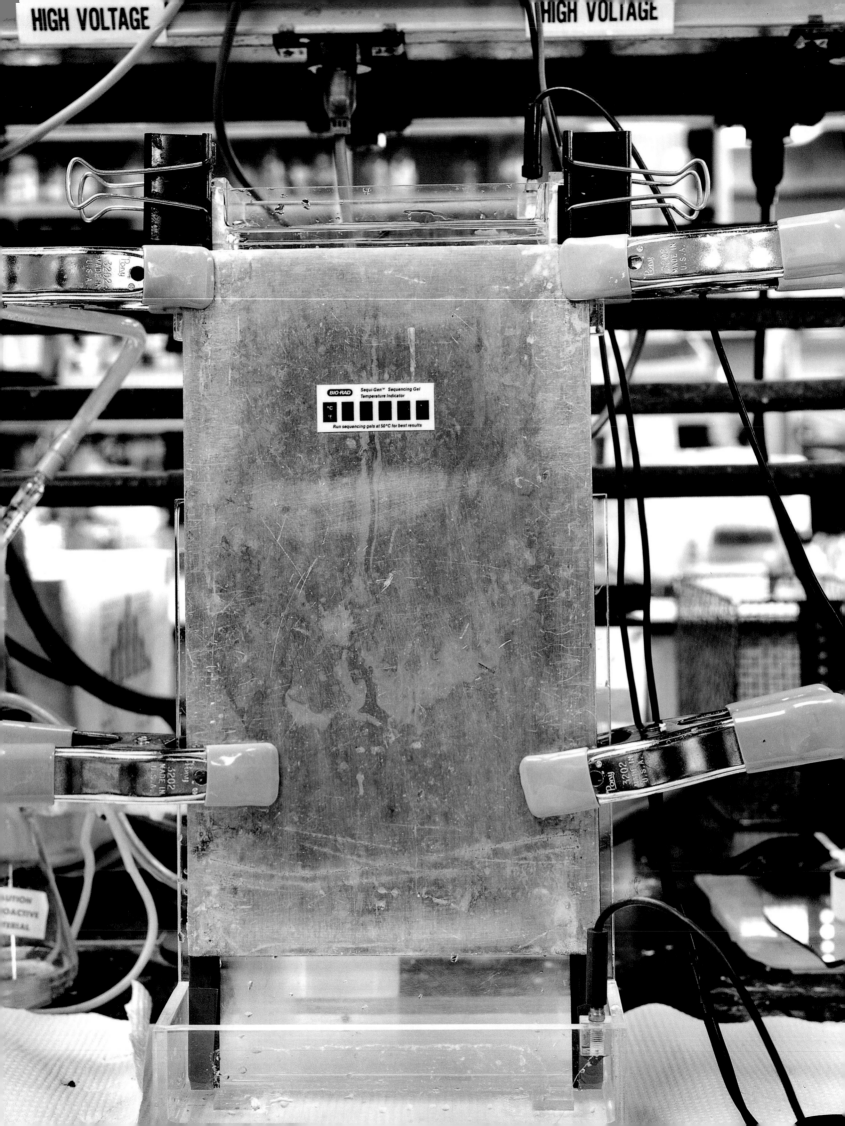

BIO-RAD Sequi-Gen™ Sequencing Gel
Temperature Indicator

°C
°F

Run sequencing gels at 50°C for best results

Checklist of the Exhibition

1.
Sequential Molecules, 1995
9 panel typology
gelatin silver prints, 20 x 24" each
5 x 6' installation
pages 25, 44, 45, 50, 64*, 65, 66, 67, 68, 69

2.
-86 Degree Freezers, 1995
12 panel typology
gelatin silver prints, 20 x 24" each
8 x 5' installation
pages 13, 35, 46, 58, 76*, 77, 78, 79, 80,
81, 82, 83, 84, 85

3.
Fossils, 1995
9 panel typology
gelatin silver prints, 20 x 24" each
6 x 5' installation
pages 12, 49, 98*, 99*, 100, 101, 102, 103

4.
*Moon Rock Found on Moon, Moon Rock Found
on Earth,* 1994
6 panel typology
gelatin silver prints, 20 x 24" each
5 x 4' installation
pages 104*, 105, 106, 107, 108, 109,
110, 111

5.
Mating Reactions of Algae, 1995
3 panel triptych
gelatin silver prints, 20 x 24" each
5 x 2' installation
pages 29*, 30, 31

6.
Phylum Mollusca, 1994
2 panel diptych
gelatin silver prints, 20 x 24" each
2' x 3'4" installation
pages 72, 73

7.
Beating Heart—Heart Chamber, 1994
2 panel diptych
gelatin silver prints, 20 x 24" each
2' x 3'4" installation
pages 11, 92*, 93*

8.
Definitely Not Sterile, 1995
gelatin silver print, 30 x 40"
page 97

9.
Pipette Stand, 1995
gelatin silver print, 30 x 40"
page 32

10.
Glove Box, 1993
gelatin silver print, 30 x 40"
page 14

11.
Ultra High Vacuum Chamber, 1992
gelatin silver print, 30 x 40"
page 115

12.
Sex Linkage, 1995
gelatin silver print, 20 x 24"
page 95

13.
Body Pattern Formation, 1995
gelatin silver print, 20 x 24"
page 75

14.
Intestinal Mount, 1994
gelatin silver print, 20 x 24"
page 89

15.
Solvent Stills, 1994
gelatin silver print, 20 x 24"
page 36

16.
Drosophila Morgue, 1994
gelatin silver print, 20 x 24"
page 71

17.
Genetically Engineered Tomatoes, 1994
gelatin silver print, 20 x 24"
page 113

18.
Bone Tissue, 1994
gelatin silver print, 20 x 24"
page 88

19.
Drosophila Stock, 1994
gelatin silver print, 20 x 24"
page 39

20.
Library Cloning, 1994
gelatin silver print, 20 x 24"
page 57

21.
Unsterilized Dispensing Tubes, 1994
gelatin silver print, 20 x 24"
page 40

22.
Plant Sensory Physiology, 1994
gelatin silver print, 20 x 24"
page 21

23.
Radioactive Cell Growth, 1994
gelatin silver print, 20 x 24"
page 91

24.
Sterilized Pipettes, 1993
gelatin silver print, 20 x 24"
page 22

25.
DNA Sequencing, 1993
gelatin silver print, 20 x 24"
page 114

26.
Waste Vials, 1992
gelatin silver print, 20 x 24"
page 26

27.
DNA/RNA Synthesizer, 1992
gelatin silver print, 20 x 24"
page 54

28.
Bone Marrow Smears, 1993
gelatin silver print, 20 x 24"
page 87

The *Moon Rock Found on Moon, Moon Rock Found on Earth* typology is constructed from two moon rock fragments. The moon rock found on the moon was collected by the Apollo 14 astronauts in a scoop of lunar soil in 1971. Scientists date this fragment to be over 4 billion years old. The moon rock found on earth is a tiny piece of a meteorite found in Antarctica in 1989. This fragment is a piece of the moon, blasted off the moon by a large meteorite impact, probably within the last million years, then captured by the earth's gravity.

About the Artist

Catherine Wagner was born in San Francisco, California, in 1953. She received her Bachelor of Arts (1975) and her Master of Arts (1977) from San Francisco State University. She is the recipient of several major awards, including a Guggenheim fellowship, NEA fellowships, a Mellon Foundation grant, the Ferguson Award and the Aaron Siskind fellowship. She is affiliated with Fraenkel Gallery, San Francisco, and Gallery RAM, Santa Monica. She is currently Professor of Art at Mills College in Oakland, California.

SELECTED EXHIBITIONS

1977

"Cityscapes," Downtown Center Invitational Group Show, M.H. de Young Museum, San Francisco

"Urban Landscape," San Francisco Art Institute, Atholl McBean Gallery (four-person exhibition)

1978

"Catherine Wagner and Walker Evans," Equivalents Gallery, Seattle (two-person exhibition)

"Catherine Wagner's Recent Work," Simon Lowinsky Gallery, San Francisco (one-person exhibition)

"Contemporary California Photography," Camerawork Gallery, San Francisco

1981

"Large Spaces in Small Places: A Survey of Western Landscape Photography, 1850-1980," Crocker Art Museum, Sacramento

"Catherine Wagner," Orange Coast College, Costa Mesa, California (one-person exhibition)

"Catherine Wagner and Gail Skoff," University of California at Santa Cruz (two-person exhibition)

1982

"Slices of Time: California Landscape 1860-1880, 1960-1980," The Oakland Museum, Oakland

"Catherine Wagner: Moscone Site," Simon Lowinsky Gallery, San Francisco (one-person exhibition)

"Catherine Wagner," Equivalents Gallery, Seattle (one-person exhibition)

1983

"Facets of the Collection from the California Sharp School," San Francisco Museum of Modern Art, San Francisco

"Catherine Wagner," New Image Gallery, James Madison University, Harrisonburg, Virginia (one-person exhibition)

1984
"La Photographie creative," Bibliothèque Nationale, Paris

1985
"American Classroom," The Oakland Museum, Oaks Gallery, Oakland (one-person exhibition)

"20 x 24 Polaroid Exhibition," Friends of Photography, Carmel

"Photography from the Permanent Collection," National Museum of American Art, Smithsonian Institution, Washington, D.C.

1986
"Bay Area Biennial," Newport Harbor Art Museum, Newport Beach, California

"New American Photography," Min Gallery, Tokyo

1987
"New American Photography," Fogg Art Museum, Harvard University, Cambridge, Massachusetts

"Catherine Wagner, 1976-86," Min Gallery, Tokyo (one-person exhibition)

1988
"Catherine Wagner: Photographs from the American Classroom Project and the George Moscone Site," Museum of Contemporary Photography, Columbia College, Chicago (one-person exhibition)

"Changing Places: Photographs by Catherine Wagner," Farish School of Architecture, Rice University, Houston (one-person exhibition)

"American Classroom: The Photographs of Catherine Wagner," Museum of Fine Arts, Houston (one-person exhibition)

"Cross Currents/Cross Country," Camerawork Gallery, San Francisco and Photographic Resource Center, Boston

1989
"American Classroom: The Photographs of Catherine Wagner," Friends of Photography, Ansel Adams Center, San Francisco (one-person exhibition)

"Photographs 1978-1988," Fraenkel Gallery, San Francisco (one-person exhibition)

"Picturing California," The Oakland Museum, Oakland

"Theme and Variations: The Photographic Still Life," San Francisco Museum of Modern Art, San Francisco

1990
"Natural History Recreated," Center for Photography at Woodstock, New York

"Perspectives on Place: Attitudes Toward the Built Environment," San Diego State University, San Diego

"American Classroom: The Photographs of Catherine Wagner," Laurence Miller Gallery, New York (one-person exhibition)

1991
"Site Work: Architecture in Photography since Early Modernism," Photographers' Gallery, London

"Silent Interiors," Security Pacific Gallery, Seattle

"Catherine Wagner: Selections," Turner/Krull Gallery, Los Angeles (one-person exhibition)

"American Classroom," Cuesta College Art Gallery, San Luis Obispo (one-person exhibition)

"California Cityscapes," San Diego Museum of Art, San Diego

1992
"To Collect the Art of Women: The Jane Reese Williams Collection of Photography," Museum of New Mexico, Santa Fe

1993
"Home and Other Stories," Los Angeles County Museum of Art, Los Angeles; Mills College, Oakland; Center for Creative Photography, Tucson (one-person exhibition)

"Home and Other Stories," Fraenkel Gallery, San Francisco (one-person exhibition)

1994
"Kunst in Frankfurt, Ein Dialog Europa-Amerika," America House, Frankfurt/Main

"Family Lives," Corcoran Gallery of Art, Washington, D.C. (three-person exhibition)

1995
"Facing Eden: 100 Years of Landscape Art in the Bay Area," M.H. de Young Memorial Museum, San Francisco

"Landscape: A Concept," Oliver Art Center at the California College of Arts and Crafts, Oakland

"Selections from the Permanent Collection," The Museum of Modern Art, New York

"Assembled Photographs," The Los Angeles County Museum of Art, Los Angeles

SELECTED COLLECTIONS
Bibliothèque Nationale de Paris, Paris, France
Center for Creative Photography, Tucson, Arizona
Chase Manhattan Bank, New York, New York
Delaware Art Museum, Wilmington, Delaware
First National Bank of Chicago, Chicago, Illinois
Grunwald Center for the Graphic Arts, Los Angeles, California
International Center of Photography, New York, New York
J.B. Speed Art Museum, Louisville, Kentucky
Los Angeles County Museum of Art, Los Angeles, California
Metropolitan Museum of Art, New York, New York
Minneapolis Institute of Arts, Minneapolis, Minnesota
Museum of Fine Arts, Houston, Texas
Museum of Folkwang, Essen, Germany
Museum of Modern Art, New York, New York
National Museum of American Art, Washington, D.C.
Oakland Museum, Oakland, California
Polaroid Collection, Cambridge Massachusetts
Rhode Island School of Design, Providence, Rhode Island

San Francisco Art Commission Archives, San Francisco, California
San Francisco Museum of Modern Art, San Francisco, California
Security Pacific Bank, Seattle, Washington
Tokyo Institute of Polytechnics, Tokyo, Japan
Victoria and Albert Museum, London, England
Washington University Gallery of Art, St. Louis, Missouri

SELECTED MONOGRAPHS

1987
Catherine Wagner, 1976-1986. Tokyo: Min Gallery.

1988
Tucker, Anne Wilkes. *American Classroom: The Photographs of Catherine Wagner.* Houston: Museum of Fine Arts.

1990
Changing Places: Photographs by Catherine Wagner. Houston: Farish Gallery, Rice University.

1993
Conkelton, Sheryl. *Home and Other Stories: Photographs by Catherine Wagner.* Los Angeles and Albuquerque: The Los Angeles County Museum of Art and The University of New Mexico Press.

SELECTED BOOKS AND CATALOGS

1978
Contemporary California Photography. San Francisco: Camerawork Gallery.

1980
Large Spaces in Small Places: A Survey of Western Landscape Photography, 1850-1980. Sacramento: Crocker Art Museum.

1982
Slices of Time: California Landscapes 1860-1880, 1960-1980. Oakland: The Oakland Museum.

1986
Bay Area Biennial. Newport Beach, Calif.: Newport Harbor Art Museum.

New American Photography. Tokyo: Min Gallery.

1989
Roth, Moira, editor. *Connecting Conversations, Interviews with 28 Bay Area Women Artists.* San Francisco: Eucalyptus Press.

1990
Sullivan, Constance. *Women Photographers.* New York: Harry Abrams.

1991
Caiger Smith, Martin. *Site Work: Architecture in Photography since Early Modernism.* London: Photographers' Gallery.

Stofflet, Mary. *California Cityscapes.* San Diego: San Diego Museum of Art.

1992
Janis, Eugenia Parry. *To Collect the Art of Women: The Jane Reece Williams Collection of Photography.* Santa Fe: Museum of New Mexico.

1994
Family Lives. Washington, D.C.: Corcoran Gallery of Art (brochure).

Kunst in Frankfurt, Ein Dialog Europa-Amerika. Frankfurt/Main: Museum für Moderne Kunst.

1995

Facing Eden: 100 Years of Landscape Art in the Bay Area. San Francisco: M.H. de Young Memorial Museum.

SELECTED ARTICLES AND REVIEWS

1977

Fischer, Hal. "The Contemporary Landscape." *Artweek* 8, no. 8 (February 19, 1977): 11-12.

——. "Approaches to Landscape: San Francisco Bay Area." *La Mamelle Magazine: Arts Contemporary* 2, no. 4 (1977): 18.

Richards, Paul. "American Aesthetic in Plain and Spooky Places." *Washington Post,* June 18, 1977.

1978

Albright, Thomas. "Photographers with Compelling Vision. Review: Irving Penn, Marion Post Wolcott, Catherine Wagner." *San Francisco Chronicle,* May 6, 1978.

Fischer, Hal. "Contemporary California Landscape: The West is...well, different?" *Afterimage* 6, no. 4 (November 1978): 4-6.

——."San Francisco. Catherine Wagner, Simon Lowinsky Gallery." *Artforum* 18, no. 1 (September 1978): 88-89.

Welpott, Jack. "East Is East and West Is West." *Untitled (Friends of Photography)* 14 (1978): 10-13.

1982

Fischer, Hal. "San Francisco. 'Slices of Time: California Landscapes 1860-1880, 1960-1980,' The Oakland Museum." *Artforum* 21, no. 2 (October 1982): 77.

Glowen, Ron. "Urban Documentaries." *Artweek* 13, no. 5 (October 23, 1982): 11.

Thomas, James W. "What Is Not Art?" *Northwest Photography* (October 1982).

1983

Welpott, Jack. "George Moscone Site." *Picture Magazine* 20 (1983): 66-69.

1985

Bloom, John. "Catherine Wagner at the Oakland Museum." *Photo Metro* 3, no. 28 (April 1985): 14.

Fischer, Hal. "Oakland. Catherine Wagner." *Artforum* 24, no. 2 (October 1985): 129-30.

Murray, Joan. "Studying the Classroom." *Artweek* 16, no. 4 (March 16, 1985): 11.

1986

Baker, Kenneth. "A Bay Area Biennial." *San Francisco Chronicle,* November 9, 1986.

Wilson, William. "The Bay Is on View." *Los Angeles Times,* October 26, 1986.

1987

Muchnic, Suzanne. "American Photography as Seen from Japan." *Los Angeles Times Book Review,* December 20, 1987.

Murray, Joan. "Photography: Books from Exhibitions." *Artweek* 18, no. 42 (December 12, 1987): 15-16.

1988

Chadwick, Susan. "Photo Exhibition Goes to School on the Lessons." *Houston Post,* September 18, 1988.

Grundberg, Andy. "Photography." *New York Times Book Review,* December 4, 1988.

"Holiday Books." *Philadelphia Inquirer,* December 11, 1988.

Reeve, Catherine. "Two Exhibits Coax the Eye to Comprehend as Well as to See." *Chicago Tribune*, October 28, 1988.

1989
Berkson, Bill. "Catherine Wagner, Fraenkel Gallery." *Artforum* 28, no. 3 (November 1989): 160.

Bonetti, David. "A Classification of Classrooms." *San Francisco Examiner*, December 8, 1989.

McCauley, Anne. "Catherine Wagner and Photographs of Urban Change." *Afterimage* 16, no. 6 (January 1989): 14-17.

Ross, Jeanette. "Organizing Meaning." *Artweek* 20, no. 43 (December 21, 1989): 10.

1990
Coleman, A.D. "Photography." *The New York Observer*, March 26, 1990.

Glowen, Ron. "The Eloquence of Empty Rooms." *Artweek* 21, no. 42 (December 13, 1990): 14.

Levy, Ellen K. "Natural History Re-Created." *Center for Photography at Woodstock Quarterly* 11, no. 4 (1990): 4-11.

"Photography." *The New Yorker* 66, no. 8 (April 16, 1990): 19.

1991
Chattopadhyay, Colette. "A Passionate Objectivity." *Artweek* 22, no. 42 (December 12, 1991): 1, 12.

Fischer, Hal. "A Conversation with Catherine Wagner." *Artweek* 22, no. 42 (December 12, 1991): 12-13.

1992
Clifton, Leigh Ann. "'...the boundaries of strict photographic concern are no longer at issue...': A dialogue between Lewis Baltz, Andy Grundberg, Sandra Phillips, and Catherine Wagner." *Artweek* 23, no. 20 (July 23, 1992): 16-21.

1993
Baker, Kenneth. "Wagner's Photos Take Measure of Domestic Interiors." *San Francisco Chronicle*, September 7, 1993.

Bonetti, David. "Homing in on Private Spaces, Personal Lives." *San Francisco Examiner*, December 10, 1993.

Kandel, Susan. "For Catherine Wagner, Every Home Tells a Story." *Los Angeles Times*, June 11, 1993.

Porges, Maria. "Catherine Wagner: Los Angeles County Museum of Art." *Artforum* 32, no. 2 (October 1993).

Whiting, Sam. "Illuminated Windows into Private Lives." *San Francisco Chronicle*, August 25, 1993.

1996
Review article in *Zyzzyva* XII, no.2 (Summer 1996).

Review article in *See—A Journal of Visual Culture* 2, no.2 (1996).

Illustration Credits

1. Catherine Wagner, *Alfred University Science Classroom*, Alfred, New York, 1987. Collection of the artist.

2. Catherine Wagner, *Seed Germination Experiment*, Calistoga, California, 1987. Collection of the artist.

3. Théodore Géricault, *Anatomical Studies of a Horse*. Ecole nationale supérieure des Beaux-Arts, Paris.

4. Rembrandt van Rijn, *The Anatomy Lesson of Dr. Nicolaas Tulp*, 1632. Mauritshuis, The Hague.

5. Albert Edelfelt, *Pasteur in his Laboratory*, 1885. Réunion des Musées Nationaux, Paris.

6. Eadweard Muybridge. *"Daisy" jumping a hurdle, saddled*, 1885. Plate 640 from *Animal Locomotion*. 1872-85. Philadelphia, University of Pennsylvania, 1887. The Museum of Modern Art, New York. Gift of Philadelphia Commercial Museum. Copy Print © 1996 The Museum of Modern Art, New York.

7. Charles Sheeler, *Crosswalks, River Rouge Ford Plant*, 1927. George Eastman House, Rochester.

8. Werner Mantz, *X-Ray Clinic*, 1926. Metropolitan Museum of Art, Ford Motor Company Collection, New York.

9. Berenice Abbott, *Penicillin Mold*, 1946. Romana Javitz Collection. Miriam and Ira D. Wallach Division of Art, Prints and Photographs. The New York Public Library. Astor, Lenox and Tilden Foundations. Berenice Abbott/Commerce Graphics Ltd, Inc.

10. Albert Renger-Patzsch, *Stairwell*, c. 1929. Metropolitan Museum of Art, Ford Motor Company Collection, New York.

11. Hilla and Bernd Becher, *Water Towers*, 1980. Courtesy of Sonnabend Gallery, New York.

12. Catherine Wagner, *Observing Skin's Protective Role*, San Francisco, California, 1987. Collection of the artist.

About the Authors

Cornelia Homburg is Curator of the Washington University Gallery of Art and works primarily on art of the 19th and 20th centuries. She has published a number of articles, exhibition catalogs and a book, *The Copy Turns Original: Vincent van Gogh and a New Approach to Traditional Art Practice* (Benjamin Publishers, Amsterdam & Philadelphia, 1996).

William H. Gass is the author of *Habitations of the Word*, which won the National Book Critics Circle Award for Criticism in 1985; his most recent novel, *The Tunnel*, was published by Alfred A. Knopf in 1995. Gass teaches at Washington University, where he is Director of the International Writers Center and David May Distinguished University Professor in the Humanities.

Helen E. Longino is a Philosopher of Science teaching at the University of Minnesota. The author of articles in philosophy of science and feminist studies, she has written *Science as Social Knowledge* (Princeton University Press, 1990) and co-edited such volumes as *Competition: A Feminist Taboo?* (The Feminist Press, 1987), *Readings in Feminism and Science* (Oxford University Press, 1996) and *Gender and Scientific Authority* (University of Chicago Press, forthcoming).

Book design by Robin Weiss Graphic Design, San Francisco.
Type composed on a Macintosh Quadra 800 in Kaatskill and Akzidenz Grotesque.
Duotone separations by Lana Repro, Lana.
Printed and bound in Germany by Sellier Druck GmbH.

eturned on or before
ned below.